Bold They Rise

Outward Odyssey
A People's History
of Spaceflight

Series editor
Colin Burgess

BOLD THEY RISE

THE SPACE SHUTTLE EARLY YEARS, 1972–1986

David Hitt and Heather R. Smith | Foreword by Bob Crippen

UNIVERSITY OF NEBRASKA PRESS • LINCOLN AND LONDON

Manufactured in the United States of America

Library of Congress Cataloging-in-Publication Data
Hitt, David.
Bold they rise: the space shuttle early years, 1972–1986 / David Hitt and Heather R. Smith; foreword by Bob Crippen.
pages cm.
—(Outward odyssey. A people's history of spaceflight)
Includes bibliographical references and index.
ISBN 978-0-8032-2648-7 (hardcover: alk. paper)—
ISBN 978-0-8032-5548-7 (pdf (web))—
ISBN 978-0-8032-5549-4 (epub)—
ISBN 978-0-8032-5556-2 (mobi) 1. Space Shuttle Program (U.S.) 2. Space shuttles—United States—History. 3. Manned space flight—History. I. Smith, Heather R. II. Title. III. Title: Space shuttle early years, 1972–1986.
TL795.5.H58 2014
629.44'1097309048—dc23
2013047054

Designed and set in Garamond Pro
by L. Auten.

To Finn, Caden, Bethany, Nathan, Lillian, Lila Grace, Will,
Baxter, Amelia, Andrew, Peyton, Sabrina, Kean, Elliott,
Rhys, Daniel, Lainey, and millions of other children who
will be the ones to carry on the exploration of tomorrow.

They venture forth, into the spangled night
Lured inexorably by dreams;
With vision
And resolve
To go beyond the quest of yesterday

Bold they rise, these winged emissaries
To wonders transcendent;
With audacity
And faith
In the divine promise of tomorrow.

Colin Burgess, "Bold They Rise"

Contents

Illustrations

Foreword

After John Young and I made the first flight of the Space Shuttle aboard *Columbia* all those years ago, people would sometimes ask me what the best part of the flight was. I would always use John's classic answer: "The part between takeoff and landing."

Now that it's all said and done, I think that describes what the best part of the Space Shuttle program was: the part between our first launch in April 1981 and the last landing in July 2011.

There were some low points in between, particularly the loss of both of the orbiters I had the privilege to fly and their crews, but as a whole I think the shuttle has been one of the most marvelous vehicles that has ever gone into space—a fantastic vehicle unlike anything that's ever been built.

The Space Shuttle has carried hundreds of people into space and delivered hundreds of tons of payloads into space. The shuttle gave us the *Galileo* and *Magellan* probes, which opened our eyes to new worlds, and it let us not only launch the Hubble Space Telescope but also repair and upgrade it time and time again, and Hubble has revolutionized our understanding of not only our solar system but the entire universe. The shuttle carried a lot of classified military payloads early on that probably helped the United States win the Cold War.

The Space Shuttle let us build the International Space Station. The Space Station is an incredible accomplishment, a marvelous complex, but it was the Space Shuttle that taught us that we could build a complicated space vehicle and make it work very well. The Space Station would not have been possible without the Space Shuttle.

But in those early days, I think the shuttle did something else, a little less concrete but just as important. The late '70s and early '80s weren't really a great time for the United States. We'd basically lost the Vietnam War. We'd been through economic hard times, through the hostage crisis in Iran.

President Reagan was shot just before our flight on STS-1. And morale for a lot of people in the country was really low. People were feeling like things just weren't going right for us.

And that first flight, it was obvious that it was a big deal. It was a big thing for NASA, but it was a big thing for the country. It wasn't just our accomplishment at NASA; it was an American accomplishment. It was a morale booster for the United States. It was a rallying point for the American people. And the awareness may not be as high now as it was then, but I think that's still true today. I think you saw that when the shuttle made its last flight; the pride people had in what it had accomplished and the fact that a million people watched it. When I talk to people, they think space exploration is something we need to be doing, for the future of the United States and humankind.

The retirement of the shuttle was kind of bittersweet for me. I'm proud of all it's accomplished, and I'm sorry to see it end. But I believe in moving on. I'd like to see us get out of Earth orbit and go back to the moon, and to other destinations, and eventually to Mars.

John and I got to see a lot of the development of the Space Shuttle firsthand. As astronauts, we were involved from an operations standpoint, and as the first crew, John and I visited the sites where they were working on the shuttle, getting it ready to fly. We had an outstanding, dedicated team, people who really believed they were doing something important for the nation. When we finally got into the shuttle for that first flight, meeting those thousands of people gave me a lot of confidence that we had a good vehicle to fly on.

I never expected to be selected for that first flight. I thought they would pick someone more experienced to fly with John. I was excited that they picked me, and I was honored to be a part of that flight. All told, that flight was the beginning of something truly amazing, and I'm honored to be one of the thousands of people who made it happen.

Bob Crippen

Preface

When I (David) first became involved in the Outward Odyssey series, working on the Skylab volume, my coauthors and I were shown a list of proposed titles for the first eight books in the series. As authors working on our first book, coming up with a title seemed like one of the more exciting parts of the job. We were thus somewhat pleased to be disappointed with the working title the publisher had provided: "Exemplary Outpost." It was an accurate title, but it lacked the poetry of the other titles on the list—titles like *Into That Silent Sea* and *In the Shadow of the Moon*. I'm not sure that we quite lived up to that standard with *Homesteading Space*, but we made our best effort.

Even though it meant giving up the privilege of titling this volume, Heather and I were quite happy to go along with the name the publisher had suggested for this book: *Bold They Rise*. It was, quite literally, poetic, taken from the poem by series editor Colin Burgess that appears as the epigraph.

When we first read the poem, very early on in the process of writing this volume, we pictured the title as being about the Space Shuttles themselves, reflecting the poem's reference to "winged emissaries." As the book took shape, however, we realized that was no longer true; the title had taken on a new meaning for us. Rather than being about the hardware, it was about the men and women who risked their lives to expand humankind's frontiers.

And in that vein, this book owes an incredible debt of gratitude to the NASA Johnson Space Center (JSC) Oral History Project, without which it quite literally would not exist.

With *Homesteading Space*, it was relatively easy to create a book that filled a unique niche—with a few notable exceptions, such as a handful of official NASA publications and David Shayler's *Skylab*, very little had been written about America's first space station. Breaking new ground was not a particular challenge.

With this book, the challenge was a little greater. There are more books about the Space Shuttle program, so it was somewhat harder to create something unique. Most of the previous works, however, fall into one of three categories—technical volumes, which span the entire program but include none of the human experience; astronaut memoirs, which relate the human experience, but only from one person's perspective; or specific histories, which are more exhaustive but focus on only a limited slice of the program.

Based on the overall goal of the Outward Odyssey series, a new niche we could address became clear—a book relating the human experience of the Space Shuttle program, not limited to one person's story but including a variety of viewpoints and spanning the early years of the program. Originally the goal was to create a "*Homesteading Space* of the shuttle program," but it quickly became apparent that was a misdirected goal. *Homesteading* had only three manned missions to cover, and thus we could delve much deeper and more broadly in covering them. To attempt to write about the subject of this book in that manner would be to do either the subject or the reader a grave disservice; we needed to narrow our approach to create something that was both relevant and readable.

When we began reading from the JSC oral history interviews early in our research, the ideal approach for the book became apparent. Here was a wealth of first-person experience, describing in detail what it was like to be there—what it was like to involved in the design of a new spacecraft, what it was like to risk one's life testing that vehicle, what it was like to do things that no one had done before in space, what it was like to float freely in the vacuum of space as a one-man satellite, what it was like to hold thousands of pounds of hardware in one's hands, what it was like to watch friends die.

This book almost exclusively offers the astronauts' perspective on the early years of the Space Shuttle program, and, while research for the volume drew on several resources, the extensive quoted material draws heavily from the JSC Oral History Project. It's the astronauts' story, told in their own words, about their own experiences.

Bold They Rise is not a technical volume. We would love for this volume to inspire you seek out another book that delves more deeply into the technical aspects of the shuttle. There are parts of the story that we had to deal with in what seemed like a relatively superficial manner; even dedicating an entire chapter to the *Challenger* accident and the effects it had seems woefully insufficient. Entire books could, and have, been written about the *Challenger* accident. If this book leaves you wanting to know more about that incident or other aspects of the shuttle's history, we encourage you to seek out those volumes. And of course, individual astronauts have told their stories in memoirs with more personality than we were able to capture here. The subject of this book is such that it can't be covered by any one volume exhaustively, but hopefully we have provided a unique, informative, and engaging overview here.

The chronological scope of the book was also set by the publisher to fit within the Outward Odyssey series. (Another volume, written by Rick Houston, picks up the Space Shuttle story where this one leaves off.) Initially, the ending point of the book was a bit discomfiting; the *Challenger* accident seemed a rather low note on which to end a book. There were any number of successes both before and after *Challenger*. Why would one pick the lowest point of the early years as a place to end the story? But, in a very real way, it was the best possible way to turn this history into a story arc.

As astronaut Mike Mullane wrote in his memoir *Riding Rockets*,

> *The NASA team responsible for the design of the Space Shuttle was the same team that had put twelve Americans on the Moon and returned them safely to Earth across a quarter million miles of space. The Apollo program represented the greatest engineering achievement in the history of humanity. Nothing else, from the Pyramids to the Manhattan Project, comes remotely close. The men and women who were responsible for the glory of Apollo had to have been affected by their success. While no member of the Shuttle design team would have ever made the blasphemous claim, "We're gods. We can do anything," the reality was this: The Space Shuttle itself was such a statement. Mere mortals might not be able to design and safely operate a reusable spacecraft boosted by the world's largest, segmented, uncontrollable solid-fueled rockets, but gods certainly could.*

That, then, is the story of this book—a Greek tragedy about hubris and its price. It's a story of the confidence that bred some of the most amazing achievements in human history but also led to overconfidence.

But make no mistake, this book is also a love letter. Both authors of this volume were born after the end of the last Saturn-Apollo flight; the Space Shuttle is "our" spacecraft. The *Challenger* accident occurred when we were still children; it was our "where were you" equivalent of the Kennedy assassination. In our "day jobs" as NASA education writers, we wrote extensively about the shuttle, its crews, its missions, its accomplishment and ultimately its retirement. We write this with a fondness for the shuttle, even when that means telling the story with warts-and-all honesty.

It's been an honor and a pleasure to tell this story. We hope you enjoy reading it.

David Hitt
Heather R. Smith

Acknowledgments

As mentioned in the preface but bears repeating, this volume owes a great deal of gratitude to the Johnson Space Center Oral History Project, without which it would not exist.

In addition, we are grateful to the University of Nebraska Press, and in particular to senior editor Rob Taylor, for their dedication to chronicling the history of space exploration through their publication of the Outward Odyssey series and specifically through their help and support with this volume. In addition, the authors wish to express their substantial thanks to Outward Odyssey series editor Colin Burgess, who has been a loyal shepherd, a wise counsel, and a good friend during the process.

It was an incredible honor to have astronaut Bob Crippen agree to write the foreword for this volume. For David, the journey to writing this book begins in a very real way in front of a television set in 1981 watching Bob Crippen and John Young make history, and to conclude that journey with Crippen being a part of this project is a surreal bookend to the experience.

Astronaut (and *Homesteading Space* coauthor) Owen Garriott provided much assistance early in the project, making contacts and helping to get things moving, and that assistance is much appreciated. In addition, astronaut Bo Bobko was also involved in the early stages of the book and provided insight into its direction and helped open some doors. Astronauts Hank Hartsfield and Joe Kerwin and NASA legends Chris Kraft and George Mueller also provided us with material for the book.

Phillip Fox, Jon Meek, Jordan Walker, Rebecca Freeman, Lauren McPherson, and Suzanne Haggerty read early portions of this book in progress and provided feedback.

On a personal note, the authors wish to acknowledge Finn and Caden Smith, ages seven and five at the time the original manuscript was finished, for their sacrifices during deadline work on this book.

In addition, David would like to thank the following:

Heather, who for years has made my writing better and without whom I could not have written this book.

As per last time, my father, Bill Hitt, for engendering my interest in spaceflight that set me on the path to, among other things, writing this book. Jim Abbott, for giving me my first break and being a brilliant editor and a wonderful mentor and for shaping the man I am today. Holly Snow, for opening the door for my new involvement with NASA.

Owen Garriott and Joe Kerwin, for sponsoring me through Olympus and for sharing their stories, their insight, their knowledge, their expertise, and their friendship.

All of those who traveled with me on multiple road trips to Kennedy Space Center, which occasionally involved successfully watching shuttle launches.

Heather would like also to thank the following:

David, for offering me the opportunity to coauthor a book and for shepherding me through the process.

Mrs. Hughes, for seeing potential in the writing skills of a young, tenth-grade Heather and inviting her to write for the school yearbook staff, sparking an interest in writing and communication that led me down this career path. Mr. Sandy Barnard, for believing that I could write and write well whatever I put my pen to.

The *Times-Mail* in Lawrence County, Indiana, the proud home to three astronauts, including Charlie Walker, who is quoted extensively in this book, for giving me my first professional writing job and an occasional space-related assignment that made a big difference in me ending up writing at NASA and thus ending up writing this book. I was blessed to work in a community that adores its hometown astronauts and that still gets excited about spaceflight.

Starbucks locations in Huntsville, Alabama, and Nashville, Tennessee, and the Flint River Coffee Company in Huntsville, Alabama, for hospitality and tasty coffee. Portions of this book were written and edited there.

And most important, God my Father. Any writing talent that I possess is a gift from You, and You have shepherded my life and career. May You get any and all glory for this volume.

Bold They Rise

1. The Feeling of Flying

On the one hand is the idea. On the other, the reality.

Sometimes the latter fails to live up to the former. The reality of experience doesn't always measure up to the way we picture it. So often in the case of space exploration, however, it is the idea that utterly fails to do justice to the reality.

For example, countless descriptions of the Space Shuttle document its specifications to the smallest of details. But knowing that the vehicle stands 184 feet tall and weighs 4.5 million pounds fueled for launch doesn't begin to capture the experience of standing at the base of the vehicle as it towers on the launchpad.

"I wasn't intimidated by it," recalled astronaut Mike Lounge of the first time he saw the fully stacked vehicle. "Well, that's not exactly true. The first time we went down to the Cape on our class tour, my reaction when seeing the pad, at seeing the orbiter and all that, is, 'My God, this stuff's too big. It can't possibly fly.' I think that's a common reaction. I knew how big it was, but it's different when you actually see it and you're walking underneath the orbiter and all this stuff. But having gotten over that, it was kind of fun to be there with the hardware. Everyone enjoys hardware over simulations and paper."

If the vehicle itself transcends expectations, NASA's astronauts found that so, too, did the experience of actually flying aboard the Space Shuttle. Those expectations would have gradually mounted during months of mission preparation and training, but the experience would truly begin in earnest when the highly anticipated launch day arrived.

For an astronaut, that first launch day comes only after years with NASA. Since 1978 astronauts have first been selected as "candidates" and must complete an initial orientation period, replete with training in almost every aspect of the agency's work, before becoming official members of the corps.

Then there are ground assignments supporting the program in ways that have nothing to do with getting ready for a mission.

And then, finally, years after selection, there's the crew assignment. Followed by more training and preparation. There's practice on the general things that will occur during the mission, like launch and landing, to make sure everyone is ready. There's practice for all the things that theoretically could occur during the mission but shouldn't, the potential anomalies and malfunctions the astronauts have to be ready for. There's training on mission-specific tasks, the unique things each astronaut will have to do on this particular flight. There's preparation, working with the scientists or engineers or companies or countries responsible for the mission payloads to make sure that those, too, are ready to go. So when launch day finally arrives, it's a long-awaited culmination of a great deal of time and effort.

Astronaut Terry Hart recalled his launch day at NASA's Kennedy Space Center (KSC) in Florida, home of the Space Shuttle's launch complexes: "It was a clear, cool morning there and we went through the whole morning, going through the traditions of having breakfast together, and there was always a cake there for the crew before they go out. And then going into the van and realizing that all the Mercury guys went on that van, it was really a very heady experience."

For three-time shuttle veteran David Leestma, that experience of waving to people while walking out to the Astrovan, suited up and ready for launch, was a memorable moment. "We always called that the last walk on Earth," Leestma said. "There's always crowds of people there to see you in case you never come back or something. It was one of those little bits of kind of gruesome humor. And then you go out to the launchpad, and you've been through this. You've been there many times before, because you train in the orbiter a few times and you have countdown demonstration tests and things. And this time you get to the pad and there's nobody there. You go, 'Ooh.' And the vehicle is steaming and creaking and groaning and you go, 'This is for real.'"

On the launchpad, the Space Shuttle is positioned vertically, its three major components having been stacked together in the enormous Vehicle Assembly Building at Kennedy Space Center before having been rolled out—slowly—to the launchpad atop a huge crawler. Standing tallest is the orange-brown external tank. The external tank has no engines of its own but carries the liquid fuel for the launch in two separate tanks, one containing liquid oxygen and the other holding liquid hydrogen. The tanks are supercooled to maintain the fuels at the cryogenic temperatures needed to keep them in liquid state—below minus four hundred degrees Fahrenheit in the case of the hydrogen. Fully fueled, the external tank weighs about 1.7 million pounds.

On either side of the external tank is a slender, white solid rocket booster (SRB), the two of which together provide the bulk of the power for the first two minutes of the launch. Once ignited, they together provide 6.2 million pounds of thrust. Their name comes from the fact that they carry their propellant—consisting largely of aluminum mixed with an oxidizer to cause it to burn—in a solid, rubbery form.

And then there's the orbital spacecraft itself, the winged, white-and-black orbiter. Near the nose of the orbiter is the crew cabin, where the astronauts fly the vehicle and live during their mission. Farther aft is the payload bay, with its two large doors. And in the rear are the three Space Shuttle main engines, fueled by the external tank, each capable of generating a thrust of almost half a million pounds.

By launch day, the launch complex's servicing structure has been rotated back, revealing the orbiter. The shuttle is ready for its crew. The entrance to the orbiter is through a hatch in the side of the crew cabin, near the top of the vertically stacked vehicle, almost 150 feet above the launchpad.

Leestma recalled the process of boarding the vehicle via an elevator in the launch tower and a gantry arm near the top of the structure:

As usual, people don't say much in elevators. It's true whether you're in a hotel or on the launchpad. You kind of watch the numbers tick by, and instead of floors, they do everything in feet in the elevators, so you're so many feet above sea level. And then across the gantry, and when you walk across the gantry you're looking down into the flame trench. And you've been there before, but the obvious thing that's striking you is that this is for real, we're going to go. At least you hope we're going to go today. . . . You get up to the White Room, the access arm, and there's only two, maybe three people there and that's it. There's nobody else on the pad and everybody's blocked off for four or five miles away. This is for real. And it's groaning and moaning and you know that it's going to launch, and it's fueled and ready to go. It's a big bomb there, sitting on the pad. And you hope that all the fire goes down and you go up, and let's go, let's get on it with it. It's great. . . . We got strapped in, and again, the guys strapping us in were a lot of the same guys that strapped in Al Shepard on his flight [to become the first American in space during Project Mercury]. So it was a very heady time. . . . You get in and you just can't wait for it to happen.

Astronaut Jerry Ross, who was the first to launch into space seven times, said journeying out to the launchpad when the vehicle is fully fueled and ready to go is quite different than going out there any other time, not only because of the reality of the situation, but because the shuttle itself is different.

The vehicle really does give you this sense that it's an animal that's awake and just ready to go do something. When you go out there and the vehicle's not fueled, it's not hissing, it's not boiling off vapors, it's not making noises that you don't hear, that you do hear when it's fueled. And there's the tremendous amount of anticipation. My first flight was the twenty-third flight of the shuttle, and I had listened to every crew come back, and I took very detailed notes of their debriefings, which were quite exhaustive early on. I listened to everything they said, and they would give us a very detailed description of what it was like, what the sensations were of launch. I put that into my databank, and I would daydream about that when I'd go running or work out at the gym or something like that. I knew it was going to be a pretty exciting ride.

The crew cabin of the shuttle has two levels. The "upper" deck is the flight deck, where the commander and pilot sit at the vehicle's controls, with a bank of large windows in front of them. The flight deck has room for up to two more astronauts to sit during launch, and behind them are windows looking into the payload bay and the controls for the orbiter's robotic arm.

Below the flight deck is the mid-deck, where the rest of the crew sits during launch. Once in orbit, the mid-deck serves as the primary living area for the crew, with storage lockers and the orbiter's kitchen and bathroom and main sleeping area. The mid-deck also provides access to the vehicle's payload bay. During launch, the mid-deck has very limited visibility, and the astronauts sitting there depend largely on word from the flight deck and the very obvious physical sensations of launch to know what's going on during ascent.

Prior to launch, once the crew members have boarded the orbiter and been strapped into their seats, the waiting begins. Traditionally, the astro-

1. STS-1 crew members Commander John Young (*left*) and Pilot Bob Crippen inside Space Shuttle *Columbia* in the Orbiter Processing Facility at the Kennedy Space Center. Courtesy NASA.

nauts board about three hours before the scheduled launch time, lying on their backs in their chairs until launch.

Very often, this is as far as things get. Any number of issues, from unacceptable weather conditions to a technical glitch with the vehicle and more, can result in the launch being scrubbed and pushed back. In those cases, the astronauts are helped out of the vehicle, and work begins to prepare for the next launch attempt. "Probably one of the most frustrating things is when you get near your takeoff time, your launch time, and then you know there's a problem, and you go, 'Please solve it. We don't want to wait here. Get us off the pad,'" noted Leestma. "The last people you want to have to make the real technical decision whether you go or not is the crew, because they're always, 'Go.' 'Yeah, we'll be fine. Let's go.' That's why you've got a whole team of folks in the launch control room doing that."

But on other occasions, the weather does what it's supposed to, the vehicle is operating properly, any number of other factors come together as they should, and launch preparations continue to proceed. Finally, as launch nears, the Space Shuttle main engines "gimbal," or tilt, to test that they will move properly, and at five seconds before launch they are ignited to make sure all three engines are functioning properly. The vehicle continues to sit on the pad, but the firing of the engines causes it to pitch slightly. It then rocks slightly back, a process called the "twang," and when the stack is vertical again, at T minus zero, a spark at the top of the fuel casing of the solid rocket boosters ignites the propellant. With more than seven and a half million pounds of thrust pushing the Space Shuttle upward, it begins to move.

Shuttle pilot and commander Fred Gregory recalled the feeling of the main engines first firing, describing it as almost a nonevent. "You could hear it; you were aware of it. It sounded like some kind of an electric motor at some distance, but you looked out the window and you saw the launch tower there and the launch tower moved back. At least that's what you thought, but then you realized the orbiter was moving forward and then back, and when it came back to vertical, that's when those solids ignited and there was no doubt about it. You were going to go someplace really fast, and you just watched the tower kind of drop down below you."

At the very beginning of the ascent, there's the brilliant light of the engines, which no photograph or video can truly capture: a brightness that seems to puncture the sky. The brilliance of the flames from the engine is dramatic during the day, and far more so when they light up the sky at night. Payload specialist astronaut Charlie Walker recalled the experience of launching on the Space Shuttle in the dark:

At night, you look outside, and this launchpad is a blue gray from the xenon light reflections bouncing off of it, with a completely black background behind it. All of a sudden the launchpad brightens up with the solid rockets igniting. The launchpad brightens up to a yellow gray, but then the whole background, suddenly there's like a sunrise that's happened over Florida. You can see the Florida landscape for miles back that way. Sure, the sky is still black, but suddenly Florida has been illuminated by a new sunrise. I can see the Florida countryside, and it's a yellow, white-yellow-orange color, the coloration of the brilliant, hot flame from the solid rocket boosters.

Like Gregory, Jerry Ross recalled that, while he was aware when the main engines first ignited, things didn't really get exciting until the solid rocket boosters fired.

As the shuttle's main engines came up, you could really feel the vibrations starting in the orbiter, but when the solid rocket motors hit, when they ignite, it's somebody taking a baseball bat and swinging it pretty smartly and hitting the back of your seat, because it's a real "bam!" And the vibration and noise is pretty impressive. The acceleration level is not that high at that point, but there is that tremendous jolt as the solid rocket motors ignite, and you're off. I'll never forget the vibrations of the solid rocket motors. As we accelerated in the first thirty seconds or so, the wind noise on the outside of the vehicle just became really intense, like it was just screaming. It was screeching on the outside. I was already thinking about "what am I doing here" before then, but [it was] just a sheer, incredible experience of the energy.

In many ways the flight deck, with its large windows, is the superior seating for experiencing the launch. In one way, however, the mid-deck has the advantage. Since the pilot and commander are busy with the tasks of making sure the vehicle is operating properly during ascent, they don't have the luxury of stopping to really take in the experience of the launch. While the astronauts on the mid-deck don't have the same view as those on the flight deck, they have the freedom to focus more on the sensations. Hart, for example, recalled being able, as a mission specialist, to really enjoy the experience.

You talk a lot [to other astronauts about what launch is like], obviously, and you see a lot of pictures, and you think about it a lot, so you think you're pretty well prepared and you probably won't have too many surprises, but I had a couple of surprises. The shake, rattle, and roll for the first two minutes, that was about what I thought, maybe even a little bit less than what I thought it would be, because the solid rockets kind of have a "whoof-whoof" [rumble]. You don't really hear it; you more feel it. It's like a very low-frequency rumble, and just a tremendous sense of power as you lift off and all.

Another part of the experience that simply cannot be replicated on the ground is the pressure of the g-forces during ascent, according to STS-6 commander P. J. Weitz: "The value of our simulators ends when those engines light and you lift off. They try to fake you out a little bit by tipping the Shuttle Orbiter Simulator and that, but it doesn't compare with three shuttle main engines and two solids going. As I tell people, I said, 'You know you're on your way and you're going somewhere and you hope they keep pointed in the right direction, because it's an awesome feeling.'"

Weitz compared the launch of the Space Shuttle to the launch of a Saturn IB, which he took into space on the first Skylab mission. The Saturn, he said, produced about half again as much acceleration force as the shuttle's three gs, and the force was felt in somewhat different ways on the two vehicles. In the Saturn, the thrust was "actual," or directly in line with the vehicle, so the crew was pressed directly back into the couches. With the shuttle, on the other hand, because of the way the orbiter is stacked on the external tank, the thrust from the main engines is offset from the vehicle's center of gravity, meaning that the crew members aboard felt the pressure pushing them not only into the back but also into the bottom of their seats.

After clearing the launchpad, the shuttle begins to roll so that the orbiter is below the external tank, to better allow its engines to offset the tank's weight. Around one minute into flight, the shuttle encounters "Max Q," the period in which the increasing velocity of the vehicle produces the maximum amount of pressure on the shuttle before the decreasing resistance of the atmosphere reduces that pressure. To reduce the strains of the pressure of Max Q, the vehicle throttles down its engines and then, seconds later, past the point of maximum pressure, throttles back up.

Just over two minutes into the launch, the solid rocket boosters separate from the vehicle, and the orbiter and external tank continue toward orbit. The solids deploy parachutes and land in the ocean, where recovery ships locate them and bring them back for refurbishment and reuse.

"At the solid rocket motor separation . . . there was this brilliant orange flash, orangeish-yellow flash across the windscreen, and then the solid rocket motors are gone," Ross recalled. "As the solid rocket motors tailed off, like at a minute forty-five or so, it almost felt like you had stopped accelerating, almost like you'd stopped going up. At that point we were already Mach 3-plus and well above most of the sensible atmosphere at that point, some twenty miles high or so. And at solid rocket motor jettison, then you're at four times the speed of sound and twenty, twenty-five miles high."

Hart also recalled the separation of the solid rockets as a memorable experience. For the first two minutes of ascent, the g-forces that the crew experiences have been building up, and then, at SRB separation, they drop off dramatically.

Very quickly, then, the solid rockets taper off and separate, and that was the first surprise I had. . . . The sensation that you have at that point I wasn't quite prepared for, because you go from two and a half gs back to about one and a half. Well, when you get used to two and a half, and it feels pretty good. You're going somewhere, you know. When you go back to one and a half, [it] feels like about a half. So you don't think like you're accelerating as much as you should be to get going. And, of course, I had worked the main engine program anyway, so I was very familiar with what the engines could do or not do. And I think in the next minute, every five seconds I checked the main engines to make sure they were running, because I swear we only had two working, because it just didn't feel like we had enough thrust to make it to orbit. But then gradually the external tank gets lighter, and as it does, of course, then, with the same thrust on engines, you begin to accelerate faster and faster. So after a couple of minutes I felt like, yes, I guess they're all working.

Ross also had the experience of worrying that all main engines were not working when they actually were.

I literally had to look to see that the three main engines were still working, because it became so smooth, and it almost felt like you weren't going anywhere;

you weren't accelerating at all. . . . At one point I can remember looking back behind me out the overhead windows again. In artists' renditions of the flames coming out of the three main engines, it's a nice, uniform cone of fire back there and stuff. Not true. The fire was all over the place. It was not static. It was dancing. It was not uniform. And again you go, "Is this thing working okay?" You don't know what to expect.

As the shuttle nears the end of its powered ascent, with the bulk of the atmospheric drag behind it, it begins to accelerate dramatically. "As we got up to about the seven-and-a-half-minute point, then, is when you get to the three gs of acceleration, and that's a significant acceleration," Ross said.

It feels like there's somebody heavy sitting on your chest, and it makes it pretty hard to breathe. I mean, you kind of have to grunt to talk, and you're just waiting for this three gs to go away. . . . You're accelerating at 100 feet per second, which is basically like going from 0 to 70 miles per hour every second. So it's pretty good. And then at the time that the computers sense the proper conditions, the main engines . . . shut off and you're in zero g. And for me, the first flight, sitting in the back seat, I had the sensation of tumbling head over heels, a weird sensation. And it was the three-g transition, from three gs to zero gs. . . . But as soon as I got out of the seat, then I was okay.

The main engine cutoff, or MECO, comes around eight and a half minutes into the launch, and shortly thereafter the external tank separates from the orbiter and reenters Earth's atmosphere. As the only major component of the shuttle stack that isn't reusable, the external tank burns up on reentry.

Gregory explained how he felt in that moment, when the main engines cut off and he was floating in the microgravity of space: "The first indication that this was not a simulation was when the main engines cut off and we went to zero g, and though [Steven] Hawley, I think, had been attributed with this comment, it was a common comment: 'Is this space? Is this it? Is this real?' And it was an amazing feeling. I'd never sensed anything like this before. So this sensation of zero g was like a moment on a roller coaster, when you go over the top and everything just floats."

Hart described being surprised once in orbit, but unlike Ross and Gregory, not by the experience of zero g.

The zero g I was pretty well prepared for. As a fighter pilot and the experience at NASA *in the zero-g trainer, you're pretty familiar with what it feels like to be weightless. But what I wasn't prepared for was the first look out the window. You don't know what black is until you see space. I mean, I was startled with just how black it was. You don't see stars. You could barely see the moon; it's because there's so much light coming off the Earth and off the tiles of the shuttle, that there's a tremendous ambient light from all those sources, so your eyes are constricted greatly. And then because of that constriction, when you look into space you can't see the stars or anything. I mean, it's like really black. It's palpable. You think you can almost reach out and touch it. I don't know quite how to describe it. It's sort of like black velvet, but it's just totally palpable. . . . I guess I knew that I wouldn't be able to see the stars when we were on the day side of the Earth. But still, when you look out there and see the blackness, it really was striking to me.*

While most astronauts report experiencing an overwhelming excitement or elation upon their first arrival in orbit, Fred Gregory jokingly recalled an odd bit of disappointment stemming from his first ascent. "Since we had trained constantly for failures, I anticipated failures and was somewhat disappointed that there were no failures, because I knew that any failure that occurred, I could handle. It was where I slipped back into an ego thing. I anticipated failures that I would correct and then the newspaper would say, 'Gregory Saves Shuttle,' but heck, none of that happened. It just went uphill, just as sweet as advertised."

As pilot, Gregory said his main job once the vehicle was on orbit was to make sure it was working properly. Since there were no major issues, he found that he had frequent opportunities for looking out the window. Said Gregory,

You immediately realize that you are either a dirt person or a space person. I ended up being a space person, looking out in space. It was a high-inclination orbit, so we went very low in the southern hemisphere, and I saw a lot of star formations that I had only heard about before and never seen before. I also saw aurora australis, which is the southern lights. I was absolutely fascinated by that. But if you were an Earth person, or dirt person, you were amazed at how quickly you crossed the ground; how, with great regularity, every forty-five min-

utes you'd either have daylight or dark; how quickly that occurred, about seven miles per second; how quickly you crossed the Atlantic Ocean.

Although he was a self-described "space person," Gregory still enjoyed occasionally gazing down on Earth below him and found it a fascinating experience.

The sensation that I got initially was that from space you can't see discernible borders and you begin to question why people don't like each other, because it looked like just one big neighborhood down there. The longer I was there, the greater my "a citizen of" changed. The first couple of days DC was where I concentrated all my views, and I was a citizen of Washington DC. I was confused because I thought everybody loved DC, but [Bob] Overmyer was from Cleveland [Ohio], and Don Lind was Salt Lake [City, Utah], and Norm [Thagard] was Jacksonville, Florida, and Lodewijk [van den Berg] was the Netherlands, and Taylor Wang was Shanghai [China], so each had their own little location for the first couple of days. After two days, I was from America, looked at America as our home. Taylor, China. Europe for Lodewijk. And after five or six days, the whole world became our home.

During his flight, Gregory developed a sense not only of Earth as a whole being his home but of just how interconnected the global community truly is, and the extent to which all people are sharing one planet.

You could see this kind of sense of ownership and awareness. We had noticed with interest the fires in Brazil and South Africa and the pollution that came from Eastern Europe, but it was only with interest. After five or six days, then it was of concern, because you could see how the particulates from the smokestacks in Eastern Europe, how that circled the Earth and how this localized activity had a great effect. When you looked down at South Africa and South America, you became very sensitized to deforestation and what the results of it was with the runoff, how it affected the ecology. Then you'd have to back up and say, well, this is not an intentional thing to destroy; this is something that they use coke as part of their process, and in order to get coke, you've got to burn. So you began to look at things from different points of view, and it was a fascinating experience. So that was the science that I was engaged in, but never anticipated it. And it was a discovery for me, so as each of these other great scientists who were with us discovered something that they had never anticipated, I also did, and I think the whole crew had.

In order to live and work in space during their missions, astronauts must learn to adapt to the microgravity environment, and that adaptation varies from individual to individual. Part of the adaptation is simply learning to get around; moving through the vehicle without gravity is an entirely different process than walking through it on the ground. For many other astronauts, adaptation involves a physical unease as the vestibular system adjusts to the lack of the orienting influence of gravity.

While it may take different amounts of time for astronauts to be back to 100 percent, most are at least functioning fairly quickly, Gregory said.

Whatever the adaptation was, within a day, everybody had adapted to it and so it was just a matter of working on all the programs and projects of the projects that you had. The body very quickly adapted to this new environment, and it began to change. You could sense it when you were on orbit. You learned that your physical attitude in relation to things that looked familiar to you, like walls and floors and things like that, didn't count anymore, and you translated [from thinking about] floors and ceilings and walls to [thinking] your head is always up and your feet are always down. That was a subconscious change in your response; it was an adjustment that occurred up there. You also learned that you didn't go fast, that you could get from one place to the other quickly, but you didn't have to do it in a speedy way. You always knew that when you started, you had to have a destination, and you had to have something that you could grab onto when you got there. But, again, this was a transition that occurred, perhaps subtly, but over a very short period of time. I can remember we all kind of joked up there that we had become space things, and we were no longer Earth things anymore. The first couple of days, a lot of bloated faces, because there was no gravity settling of the liquids. But after a couple of days, you lost that liquid in your body, and you looked quite normal. So it was a fascinating experience. I think it was surprising to us how quickly we adapted to this microgravity environment.

With launch complete and their bodies adapting to space, the astronauts would go about their mission, spending days on any variety of different tasks carried out by shuttle crews during the early years of the program. Finally, though, the time would come to return to Earth. The orbiter would turn backward relative to its velocity and fire its engines to slow itself down, before rotating back to begin its descent.

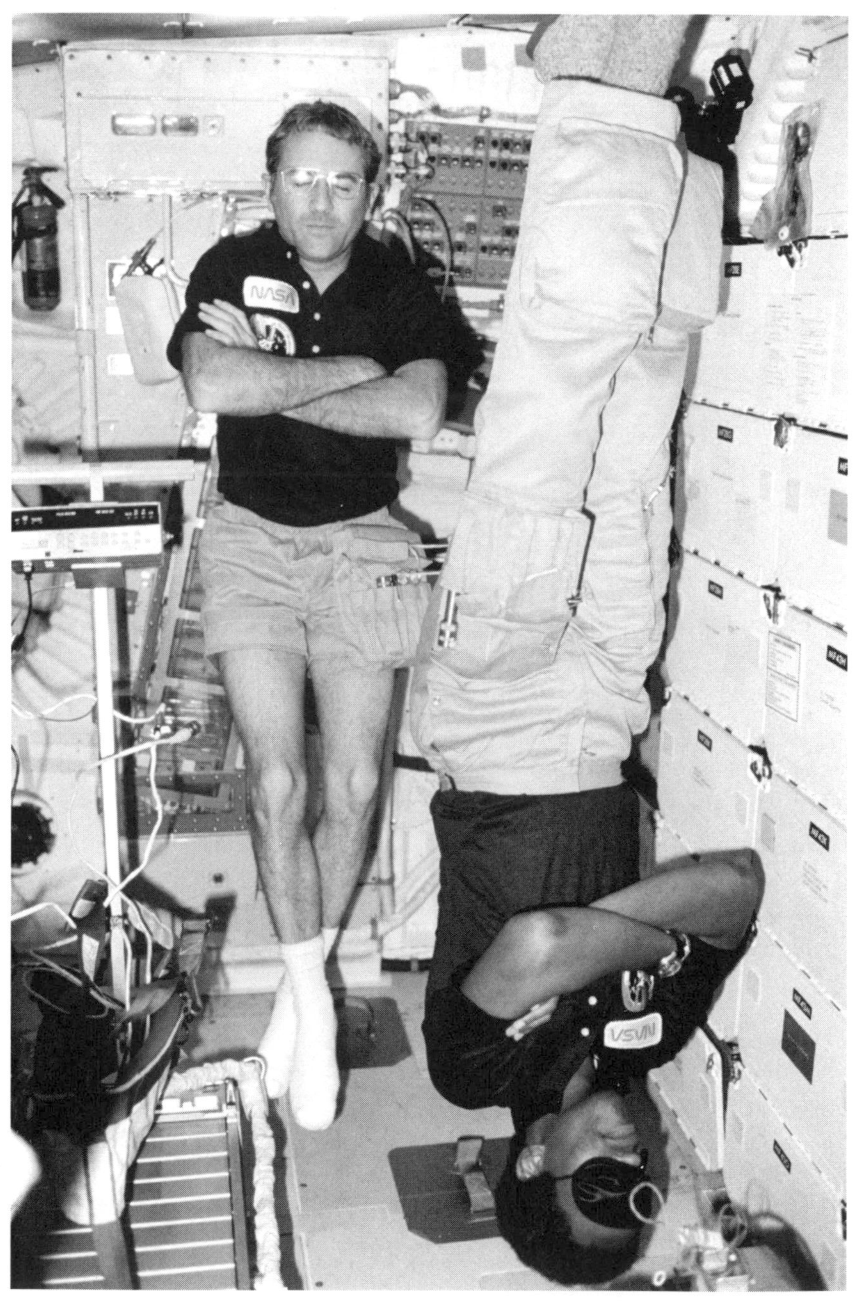

2. During STS-8, Commander Richard "Dick" Truly and Mission Specialist Guion Bluford sleep on *Challenger*'s mid-deck. Courtesy NASA.

The experiences of launch and landing are very different, Gregory said. Ascent is relatively quick and marked by rapid changes in the g-forces experienced by the crew. Landing, on the other hand, is far more gradual.

On reentry, it is entirely different. Though it takes eight and a half minutes to get up to orbit, it takes more than an hour to reenter, and it feels very similar to an airplane ride that most people have been on. You get an excellent view of the Earth. If it's night when you reenter the atmosphere, then you see a kind of a rolling plasma over the windows. . . . But other than the onset of g that occurs at less than 400,000 feet above the Earth, it is like flying in an airplane. The sensations that you have are very similar to a normal domestic airplane flight. You're going pretty fast, but you are not aware of it because you're so high.

It's an amazing vehicle, because you always know where you are in altitude and distance from your runway. You know you have a certain amount of energy and velocity, and so you also know what velocity you're supposed to land, and you can watch this amazing electric vehicle calculate and then compensate and adjust as necessary to put you in a good position to land. We normally allow the automatic system to execute all the maneuvers for ascent and for reentry, but as we proceed through Mach 1, slowing down for landing, it is customary for the pilot, the commander, to take command of the orbiter and actually fly it in, using the typical airplane controls. But, you know, as I look at it, the ascent is very dramatic. It's very fast, a lot of movement, but quick. The entry is more civilized but exposes the orbiter to actually a greater danger than the ascent, as far as the influence of the atmosphere on the orbiter. The temperatures on the outside of the orbiter really get hot on reentry, and that's not the case on ascent.

Astronaut Charlie Bolden flew on the Space Shuttle four times, two of those as commander, and became NASA administrator in 2009. A former naval aviator, Bolden described landing the Space Shuttle as a unique experience. "The entry and landing is unlike almost anything you ever experience in any other kind of aerospace machine because it's relatively gentle," Bolden said.

In terms of g-forces and stuff like that, it's very docile. Unless you do something wrong, you don't even get up to two gs during the reentry, the entire time of the reentry. When you bank to land, you come overhead the landing site, and then you bank the vehicle and you just come down like a corkscrew. . . . It feels like you've got gorillas sitting on your shoulder because you've been weightless for x number of days. And so it's just a really different feeling. You have to hold your head up because you've got this big old heavy helmet on and it probably weighs [a few] pounds, but it feels like it weighs a hundred. It takes a little bit of energy to get your hands up off the console, because once you start feeling gravity again, your hands just kind of go down and they want to stay there; everything does. So the two pilots on board are doing a lot of isometric exercises all the way down.

Even when an astronaut lands the shuttle for the first time after a mission, Bolden said, it already feels very familiar because of all the training in preparation for the missions. "It's like you've done it all your life, because you have," Bolden said.

You've done it thousands of time by now in the shuttle training aircraft for real, and you've done it probably tens of thousands of times in the simulator. So it doesn't look abnormal at all; it's just something that you're accustomed to. When you touch down, if you do it right, again, you hardly know you touched down. As big as the orbiter is, the way that we land it is we just get it into an extremely shallow approach to the landing, and so it just kind of rolls out on the runway, and if you do it right, you all of a sudden notice that things are starting to slow down real quick and you're hearing this rumble because the vehicle's rolling down the runway on this grooved runway. So you know you're down, put the nose down and step on the brakes and stop. That's it. And then you go, "Holy G. I wish it hadn't been over so quick." I don't think it makes a difference how long or how short you've been there, it's over too quick. You're ready to come home, but once you get back, you say, "Boy, I wish I had had a few more days," or something like that. And for me, my last two, being the commander and actually being the guy that had the opportunity to fly it to touchdown, was thrilling.

Once the landing is completed and the orbiter is safely back on Earth, the crew begins the process of reacclimating to the planet's strong gravity after days of feeling weightless. Charlie Walker, the first commercial payload specialist, who flew on the shuttle three times, recalled waiting in the orbiter at the end of the mission.

The guys on the flight deck were going through the closeout procedures. Ground crews were closing in. We sat unstrapped, but we would sit in our seats for another ten, fifteen minutes as the ramp was brought up, the sniffers checked for am-

monia leaks and/or hypergolic propellant leaks, found none, and put the stairway [up to the hatch], and opened the hatch. All that time, all of us are beginning to get our land legs back, unbuckle, start to try to stand up. "Ah, this doesn't feel good yet. Wait a little bit longer." So you kind of move around, move your arms first, your feet first, your legs first, then stand up, make sure you've got your balance back. The balance is the one thing that you just don't have. Again, the brain hasn't been utilizing the inner ear or senses of where the pressure is on the bottom of feet, for instance, to use as cues to balance itself against gravity. It hasn't done that for a week. So you've got to carefully start through all that and consciously think about balance and consciously think about standing up, and we very consciously do that, because the last thing you want to do, in front of hundreds of millions of people watching on television, is to fall down the ramp leaving the orbiter.

Normally, on Earth, the body works hard to make sure the brain is adequately provided with blood. From a circulatory perspective, the brain, the part of the body that most needs blood, is located inconveniently at the top of the body, so the heart has to pump blood against gravity to get it there. In orbit, on the other hand, blood flows much more easily to the head, but it doesn't fill the legs the same way without gravity pulling blood into them. Astronauts develop bloated heads and "chicken legs" due to the body's confusion over how to distribute blood without gravity. The body takes the increased fluid flow to the head as a sign that it is overly hydrated and begins to shed what it sees as excess fluid. After the return to Earth, fluid redistributes again, which can cause problems.

"The body adapts by, among other things, letting go of a lot of fluid, about a liter of liquid, which makes you clinically dehydrated while you're in space, except the whole condition of the body is different up there, so you're really not dehydrated in that environment," Walker said.

But if you come back without replacing that liter of fluid, then you are dehydrated. You try to stand up with not so much fluid to go to the head, and so you literally could pass out. Nobody did that, but I know I had sensations of lightheadedness for the first few minutes until I just literally worked at getting my balance back and focusing attention, and the body was adapting all that time, too. But leaving the spacecraft, I was holding onto the handrail as I went down the stairway. Got to the bottom of the stairs, and I was walking like a duck, because I was trying to keep my balance.

Once they've adjusted enough to walk, crew members board the Astrovan, which takes them to the medical quarters for postflight medical exams and a shower. Walker said it felt good to take a shower after days without one.

Every sensation for the next many hours, normal sensations of water running over you in a shower, [felt] strange. Because again, here this water's hitting you, and it's running down. And hours later, I found that I still could at any moment just think about the sensations in my body, and it was odd to feel this pull down toward the surface of the Earth, to be stuck to the surface of the Earth. [When I flew], it was still fairly new to hear comics or some wag note that this or that "sucks." [Coming back] the astronauts were saying, "Well, the Earth really does suck." So it keeps me drawn right down to the surface. Gravity is really real, and it stands out in your mind to, again, the freedom of weightlessness when you've had that opportunity. And that was just very much on my mind. I remember even a day, two days later, probably like a day later at a meal, I was sitting down, and I could not easily figure out whether I should sit back against the back of the seat or lean forward, because my head was telling me I was leaning forward at an angle, and, in fact, I was sitting almost straight up and down. So the inner ear is still adapting to its own senses and the body's cues to orient itself and still doesn't have itself figured out completely yet.

Even if an astronaut spends only a few days in orbit after a lifetime of living in the gravity of Earth, habits developed during those few days of weightlessness can persist for a little while after the mission. "I also remember waking up the next morning back here in Houston, waking up and going into the bathroom and wanting to brush my teeth, and I did that, and I remember letting go of the toothbrush, and it fell to the sink top, and I probably laughed," Walker noted.

Then I pick up the cup of water to rinse my mouth out, and then proceed to let that cup go again. It's like, again, you're still thinking weightlessness, and you're really used to that. Finding the situation where gravity is ever-present is just such an interesting experience, because now, again, you've had that contrast of a different place where that wasn't part of the environment and you note when you get back how remarkable and how constraining gravity is. . . . We've all grown up for some decades, before we go fly in space, in gravity, and it's just natural. Except it is programmed in, and that programming is submerged with new habits that you gained to work in weightlessness, and you have to pull that programming back, or the brain does, and it does so at different rates, I think. So within tens of minutes, you can walk comfortably. You may look a little odd, because you're not walking as expertly as you had done for twenty, thirty, forty years before. It takes a few more hours, maybe a couple hours to do that. But you can walk, so balance comes back pretty darn quickly. But it's probably the nonautomatic stuff, like I've remarked about just automatically leaving a glass hanging in the air, thinking it's going to stay there. You just get into habits there that are semiconscious, and it takes a little while for the body and the brain to let go of that and to relearn that, no, I'm stuck here again to the surface of the Earth. I've got to put the glass right up here on the table directly.

2. In the Beginning

Arguably, it could only have happened when it did.

Astronaut John Young, who would go on to become the commander of the first Space Shuttle flight, was standing on the surface of the moon during the *Apollo 16* mission in April 1972 when he heard the news that Congress had approved vital funding for the development of the shuttle in its budget for fiscal year 1973. He reportedly jumped three feet into the air on the lunar surface upon hearing the news.

The Space Shuttle would be the most complex piece of machinery built by humankind. It was an incredible challenge and a daunting undertaking. At another point in history, a decade earlier or even a decade later, it might have seemed too challenging, too ambitious. But the project was born when men were walking on the moon. From that perspective, anything was possible.

It would be, far and away, the most versatile spacecraft ever built. But to many of the early astronauts who were involved in its creation, it was something even more fascinating—an aircraft like no other. Talk to the astronauts brought in as pilots during the 1960s, and there's a fair chance they'll refer to the orbiter as "the airplane." Many of them will talk about its development not in terms of rocket engines and life-support systems but in terms of avionics and flight control systems. They had been pilots, many of them test pilots, and they had come to NASA to help the agency fly capsules through space. But now—now they were aircraft test pilots again, helping to design an aircraft that flew far higher and far faster than any aircraft before.

Since the selection of the first astronauts, members of the corps had been involved in the development of new spacecraft and equipment, providing an operator's perspective. These were the people who would have to use the things that the engineers were designing, so it was their job to give the engineers feedback on whether the things they were designing were actually usable. For much of the time the Space Shuttle was being developed, most of the astronaut corps was grounded, with only a dozen flying between the last moon landing in 1972 and the first shuttle flight in 1981. As a result, there was plenty of opportunity for astronauts to be involved in the development of the shuttle, and they participated more in the development of this vehicle than any before.

Even so, there were some at NASA with the idea that the moon would be just the first step into the solar system, who were concerned about what the shuttle wouldn't be able to do—go beyond Earth's veritable backyard.

In January 1973 astronaut T. K. Mattingly was assigned to be head of Astronaut Office support to the shuttle program. This was around the same time that the contracts were being awarded to the companies that would be responsible for making the shuttle's various components. Mattingly, who had orbited the moon on *Apollo 16* while Young was walking on it, recalls talking to Deke Slayton, the head of flight crew operations at NASA's Johnson Space Center (JSC) in Houston, Texas, about the assignment. "When I got back from *Apollo 16*, Deke asked me, he said, 'You know, there's only one more flight, so if you really want to fly again anytime near-term, you might want to take the backup assignment on [*Apollo*] *17*,' he said. 'Chances aren't very good, but we do know that we replace people occasionally. So if you would like to have that chance, you can do it, or you could work on the shuttle program.' Really, I hadn't paid much attention to it," Mattingly said of the shuttle program at that point.

I kind of knew the work was going on, but I didn't know what it was, because my ambition had always been—I didn't think I would go to [walk on] the moon, but I was really hoping that I'd get to be on the Mars mission, which I was sure was going to happen the following year. To a young kid, it just seemed obvious that the next step is you go to the moon, then you sharpen your tools and you go to Mars, and I thought, "Boy, that's where I'd like to go."

Even by then it was becoming obvious that that wasn't really a likely proposition. I wasn't enthused about the shuttle because I still thought going to Mars was the next step. I believe that we needed to build a space station first so we could have hardware, which would gather years of lifetime experience while we could get to it and fix it, and we could build the transportation system while we're gaining the experience with a space station. All of that architecture was obviously politically driven, and they were having to fit into a tighter budget.

There really was not a great swell of emotion or enthusiasm for things following Apollo in the political arena, nor in the public arena, for that matter. So I think they had to walk some very, very tight lines in order to keep the program going, and so they chose the Space Transportation [System] as the way to go.

George Mueller, the head of manned spaceflight at NASA during the Apollo program and the man many recognize as the father of both the Skylab space station and the Space Shuttle program, said that, even with the development of the shuttle, human exploration of other worlds remained the ultimate goal. "It became clear that the cost of getting into orbit was the driver for all future programs. I began to think about, how do you get the cost down. In air travel, you can't fly from here to London and then throw the plane away when you get to London. What we came up with was a completely reusable vehicle. We had every intention of going back to the moon. What we were doing was going into low Earth orbit and establishing a base there; it was a requirement for reaching our long-term goal."

Former Johnson Space Center director Chris Kraft recalled the approval of the shuttle as "a real come-down for NASA."

We, the powers that be at NASA, had grand visions of going back to the moon, having bases on the moon, and on to Mars. They made very significant reports on what the future of NASA could and should be. But when the Nixon administration decided that the limitations of the budget in his [the president's] mind would not allow us to do those kinds of grand things in space, that's when the powers that be in NASA decided, well, what is the one thing that we need to start the next generation of spaceflight? And that is we need a cost-effective launch system. That's the first thing we need. If we're going to go into orbit and do grand things, or if we're going to put things in orbit and rendezvous and go other places, what we need is a good truck. We called it a truck, at times. And so that's how we arrived at that being the next step in the space program being a reusable, therefore fly-back vehicle. We signed a fixed-price, seven-and-a-half-billion-dollar contract to build the Space Shuttle, and that was to be provided with annual increases in the budget for inflation. We never got the first piece of inflation at any time in the history of the budget of the shuttle. They welshed on that guarantee immediately, and furthermore, they delayed the program a year and did not give us any relief on the total cost, on the total fixed cost. They didn't want the money in the budget that year, just that simple. So in the history of the shuttle program, up until we made the first flight, we were always pushing a bow wave of being behind budget.

Many in the astronaut corps had doubts as to what the shuttle decision would mean for the future of exploration. Mattingly considered leaving NASA completely, believing he would probably never leave Earth orbit again.

I went up to pay courtesy calls to the navy after we got back, and John Warner was then secretary of the navy, and we made a courtesy call to him. He was all enthusiastic. He says, "You navy [astronaut] guys need to come back, and we'll give you any job you want. You pick it. Whatever you'd like. You want a squadron? You want to do this? Just tell me. It's yours." Boy, my eyes lit up, and I thought, "Wow." One of my escort officers was a captain in the Pentagon. He went back and told his boss, who was the chief of naval aviation, what Warner had said, and very quickly I had an introduction to the chief of naval aviation, who made sure that I understood that despite what the secretary had said, in the environment we were in, I was not going to come in and take over his squadron. He'd find a place for me, he'd give me a useful job, but don't think that with the Vietnam War going on and people earning their positions the hard way, that I was going to walk in there and do that. He says, "The secretary means well, but we run the show."

So armed with that piece of information that if I went back on real navy duty at that point I was probably not going to find a particularly rewarding job, I thought the opportunity to get in on the shuttle at the beginning and go use some of the experience we gained would be useful, so I told my sponsor I'd do whatever the navy preferred I do. After all, they gave me my education and everything else that mattered. "So you tell me, but if I had a vote, I would say why don't I stay because the shuttle program's only going to take four years." That's what we were advertising. You know, four years, that's not all that long. So after a significant amount of discussion within the navy side of the Pentagon, they said, "Okay. Well, we agree. You probably can contribute more if you stay there." So that lead me to stay with the shuttle program, and so the beginning of that was a period of a great deal of the turmoil of getting started.

Step one of designing a Space Shuttle was deciding exactly what a Space Shuttle should be designed to do. Its official name, the Space Transportation System, summarized a basic part of the requirement. The shuttle would transport astronauts and cargo from the surface of Earth into space and back. It also was to be, as much as possible, reusable. The idea was that creating a spacecraft that was as reusable as possible would cut down on what had to be built for each launch, and thus on the cost of each launch. Lower the cost of putting a pound of material in orbit, and you can put more pounds of material in orbit. The space frontier opens up.

"We had a general idea of what specifications the shuttle was supposed to be, but in those days it was substantially larger and more aggressive than what we know today," Mattingly said. "So we went through this requirements refinement where everybody broke up into groups to go lay out what they had to do, and it evolved into something we called design reference missions. Rigidly, the idea was, we knew the shuttle was going to last for decades, and we knew nobody was smart enough to define what those missions that would come after we started were going to evolve into. So we took great pride in trying to define the most stressful missions that we could."

Mattingly said the program initially outlined three types of possible missions. One was for the shuttle to be used as a laboratory. "We laid out all the requirements we could think of for a laboratory—the support and what the people need to work in it, and all that kind of stuff," Mattingly recalled. A second type of mission was defined as deploying a payload on orbit. "That was to be one that launched and had the manipulator arm and cradles and all of the things necessary to do that."

Then there was the idea of a polar mission. Such a mission would involve putting the shuttle in a polar orbit—leaving the launch site and heading into a north–south inclination that would cause it to orbit from one pole to the other. A satellite in polar orbit would be able to fly over any point on the surface of Earth—a valuable capability for intelligence gathering. "The polar mission was really shaped after a DoD [U.S. Department of Defense] requirement," Mattingly said.

The original mission, as I recall, was a one-rev mission. [A "rev" is essentially one orbit around Earth.] You launched, got in orbit, opened the payload bay doors, deployed a satellite, rendezvoused with an existing satellite, retrieved it, closed the doors, and landed. And this was all going to be done in one rev or maybe it was two revs, but it was going to be done so that by the time anyone knew we

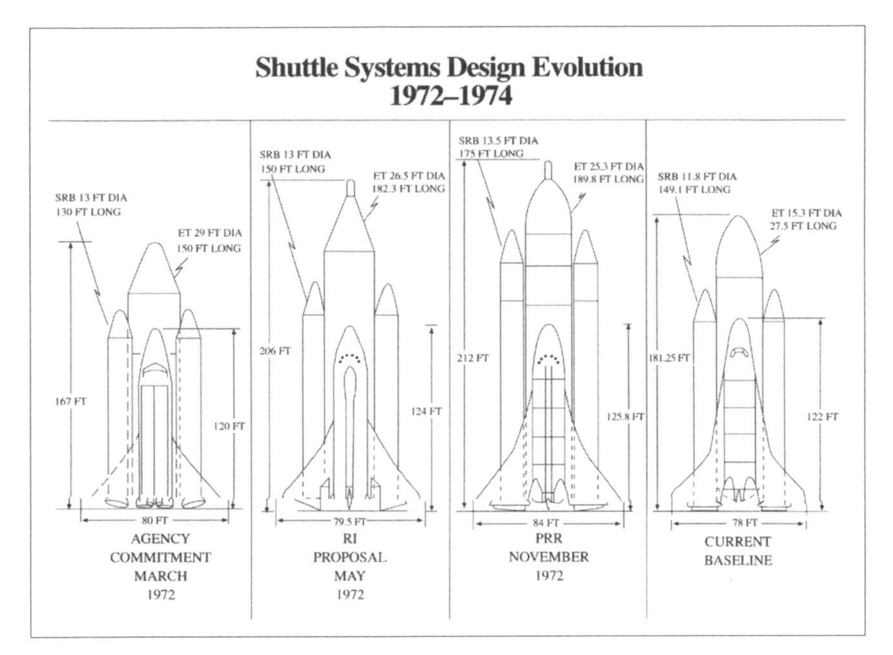

3. Space Shuttle design evolution, 1972–74. Courtesy NASA.

were there, it was all over. Well, we worked on that mission and worked on it and worked on it, and finally it became [two different design reference missions]. We just couldn't figure out how to do it all on one short timeline.

The military design reference missions were a response to a political exigency NASA had learned to deal with during the 1970s. Most notably, in developing the Skylab space station, NASA found itself competing for funding against the air force, which was seeking money at the same time for its Manned Orbiting Laboratory program. Although the two programs were very different in their goals, they shared enough superficial similarities that Congress questioned why both were necessary. With the shuttle, NASA hoped to avoid a repeat of this sort of competition, and have an easier sell to Congress, by gaining buy-in for the idea from the military. According to astronaut Joe Allen,

Leadership in the early 1970s decided the only way the Apollo-victorious NASA would be given permission to build a reusable space transportation system is that there be identified other users for the system other than just the scientists. This nation's leadership identified the other users as the military. The Space Shuttle would be used to carry military payloads. The military has its responsibilities, and they said, "All right. If our payloads are going to go aboard, we do have one require-

ment; that is that your Space Shuttle be able to take the payloads to orbit, put them there, and land back at the launch site after making only one orbit of the Earth."

The need for quick, polar missions greatly affected the design of the shuttle, yet interestingly the Space Shuttle never flew a polar-orbit mission. "At face value, that doesn't seem all that difficult to do," Mattingly said of the polar-orbit missions,

but what it meant was, the shape of the orbiter went from being a very simple lifting body-type shape, with very, very small wings, to a much larger vehicle with delta-shaped wings. I don't know the exact numbers, but the wings that go to orbit and come home again [make up a large portion of] the weight of the vehicle, and they're never fully used; only the outermost wingtips are used. All that vast expanse—with all that tile, and all the carbon-carbon [carbon-fiber-reinforced carbon] along the leading edge—is never used. It would be used if it were to go to space in a polar orbit and then come home. It would be used to gain the fifteen hundred miles of cross-range that one needs because the Earth moves fifteen hundred miles in its rotation during the time you've gone once around. So you have to have some soaring ability. That's what these large wings are for. The Space Shuttle would have cost much less money. It would cost much less to refurbish each time. Still, it would not be an economic wonder, but it would be economically okay, were it not for these huge wings. Of course, that requirement, in hindsight, was never used, was never needed, but the current Space Shuttle will forever be burdened with these wings.

Mattingly also said that the design missions established the capabilities that the Space Shuttle system would need to have. Each specification let to a variety of trickle-down requirements, and gradually the vehicle began taking shape.

These requirements we set really had some interesting things. Some of them were politically defined, like you'll land at any ten-thousand-foot runway in the world. That's all it takes. In selling the program, they had to appeal to just every constituency you could find to cobble together a consortium of backers that would keep the program sold in Congress. People don't recognize how that ripples back through a design into what you really get, and, of course, by the time you know what you've got, the people who put those requirements in, they're history. So it's interesting. But that ten-thousand-foot runway requirement set a lot of limits on aerodynamics and putting wings on the airplane. The cross-

range—that was the air force requirement for this once-around polar mission abort—that sized the wings and thermal conditions. That precluded us from using a design called a lifting body that the folks out at Edwards [Air Force Base, California] had been playing with and had demonstrated in flights. It was structurally a much nicer design, but you just couldn't handle the aerodynamic characteristics that were required to meet these things. So we had a vertical fin on this thing and big wings, and it's a significant portion of shuttle's weight, and the maintenance that goes with it is attributed to the same thing.

Mattingly had the unique vantage point of watching the shuttle program evolve from a concept through logistical support into its mature state, he recalled. "I look back and I say, 'Well, we know what we started to do, and we know what we have, and they're not always the same. Why?' Because it was an extraordinary job. Apollo was a challenge because it was just so big and it was audacious, and time frame was tight, and all of those things." But in many ways, Mattingly said, the shuttle was even more challenging.

Essentially, it was so demanding that all of the engineering and ops [operations] people . . . generally stayed on. We didn't have a lot of technical attrition after Apollo. At least that's my impression. At least the middle-level guys all stayed, and they kept working it because they recognized that the shuttle was a far more challenging job than Apollo in many technical senses.

The part of the shuttle that was different was Apollo was a collection of boxes. If you had a computer, you could build it, you could test it, you could set it out and do it all by itself. You had a second stage. You could build and test the whole thing by itself. Well, with the concept of this reusability and integration, you didn't have anything until you had everything. There was no partial thing. There was nothing that was standalone. I remember we were trying to buy off-the-shelf TACANs *[Tactical Air Control and Navigation systems], an airplane navigational system, and as part of this integration process, rather than take the* TACAN *signal that an airplane generated in those days and used for navigation, we stripped it all out and put in all our own software so that this off-the-shelf* TACAN *box was absolutely unique. There was nothing else. And it was part of the philosophy of how we built this system.*

Despite the areas where the shuttle fell short of the original requirement-based specifications, Mattingly said NASA ended up with a very robust and versatile vehicle because of how ambitious the original discussions were. "At

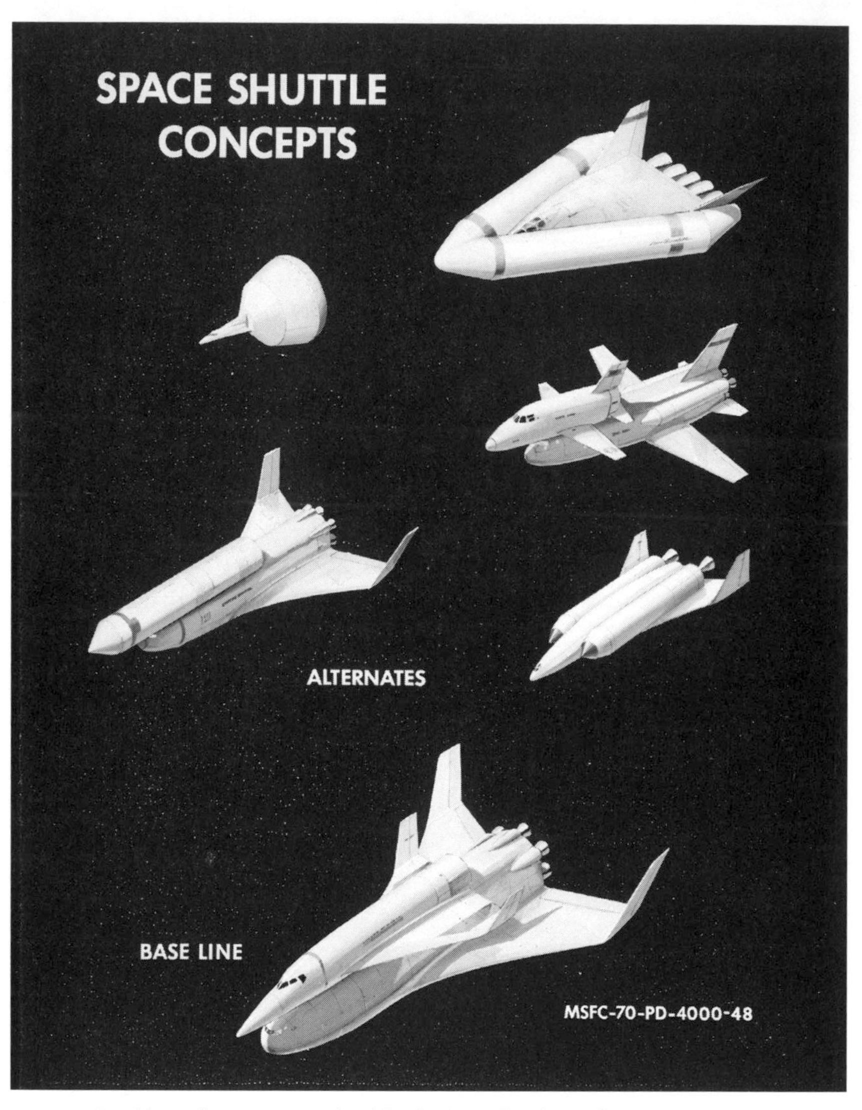

4. Possible configurations considered for the Space Shuttle, as of 1970. Courtesy NASA.

the time we were doing this and putting all these requirements on there, we were actually, I think, quite proud of having had the foresight to look at all of these things. Today you can hardly think of a mission . . . you'd like it to do that it can't do. It is an absolutely extraordinary engineering piece, just unbelievable. The shuttle really did fulfill almost all of the requirements that we were tasked to put into it."

The shuttle went through a variety of widely different configurations during its early development. An inline version would have had the orbiter on top of a more traditional rocket booster, which would use parachute recovery to make it reusable. Another version would have had the orbiter launched atop essentially another space plane that would fly back to a ground landing site.

Discussions were held as to whether the primary fuel tank, which ended up being the external tank, should be inside the orbiter or not. There were trade-offs, according to Chris Kraft, the Johnson Space Center director at the time. Putting the tank inside the orbiter would have required that the orbiter be much larger but would have greatly increased the reusability of the shuttle system. However, Kraft said, the ultimate limitation was the difficulty of designing an integrated vehicle that wouldn't suffer substantial damage to the fuel tank during landing.

Another major issue that had to be figured out early on was what sort of escape system should be provided for the crew. The Mercury and Apollo capsules both had powerful solid rocket motors in the escape towers at the top of the vehicle that would have been capable of lifting the spacecraft away from the booster in case of an emergency. "From the get-go, we tried desperately to put an abort system on the shuttle that would allow us to abort the crew and/or the orbiter off of a malfunctioning solid rocket or malfunctioning SSMEs [Space Shuttle main engines]," Kraft said.

Originally we tried putting a solid rocket booster on the ass end of the orbiter, and the more we looked at that, the more we could not come up with a structural aerodynamic qualification and weight that would accomplish that job. We looked at putting a capsule in the structure of the crew cabin, making it something that would separate. We looked at the possibility of putting a capsule in the orbiter, at the structural problems of attaching a capsule, getting rid of the front end, making it strong enough, making it aerodynamically sound, building a control system that would allow it to descend under any and all Mach numbers. And we decided if we do that, we can't build a Space Shuttle. We can't afford the mass, and we don't think we could build it in the first place.

So the answer to that question was, we will use the solid rockets that we have as our escape system and fly the orbiter back to the launch site if we have an abort. So we said to ourselves, the solid rockets have to, once you release them from the pad, bust those bolts on the pad, it has to be 100 percent reliable. And

5. An early depiction of the Space Shuttle identifies major components as the orbiter, the three main engines, the external tank, and the two solid rocket boosters. Courtesy NASA.

we always assumed it was, and any decision we made could not screw around the reliability of those solids. So our abort system was the solid rockets and a return to launch site, RTLS; you could fly that orbiter back.

Kraft said critics gave NASA a hard time about the shuttle not having an escape system. "I always thought that was unfair as hell," Kraft said. "I don't think they understood the system. And if you ask them over there [at NASA] today, I guarantee you won't find five people who understand that's what we did. But we did have an escape system, we had the solid rockets and the fly-back capability. Now, it didn't save the *Challenger*, and nothing would have saved *Columbia*. But those two accidents were created by the fallacies of man, not by the machines."

Recalled astronaut Charlie Bolden of the RTLS abort:

While a lot of us flew a lot of them in [simulation], I'm not sure any of us ever believed that that's something you really wanted to do. This was a maneuver in which something goes wrong shortly after liftoff, and you decide you're going to turn the vehicle around and fly it back to the Kennedy Space Center. And the computer's got to do that, so the software really has to work. It's crazy, because you're going upside down outbound, and all of a sudden you decide you're going to go back to Kennedy. And while you're still flying downrange, you take this vehicle and you pitch it back over so that it's flying backwards through its own fire for several minutes. What has to happen is the computer has to calculate everything precisely, because it's got to flip it over, have it pointing back to the Cape while it's flying backwards, so that just before the solid rocket boosters burn out, it stops the backwards downrange travel and starts it flying back to the Cape. And then once that happens, then the solids cut off. They separate; they go their way, and then you fly back for a few minutes, for another six minutes, and the main engines cut off and you separate from the external tank. And that became a very tricky maneuver, because what you're worried about was re-impacting with the tank, and if you did that, you were dead. So it's a maneuver that . . . nobody ever wants to fly it, because just, it's like, boy, this is really bad if you have to do this.

Once the general requirements were outlined from the mission baselines and the general type of vehicle was decided, work began on figuring out how exactly to design a spacecraft that would meet the requirements. Making the process particularly interesting was the fact that the shuttle was a collection of very diverse elements that had to be designed to work as an integrated system. The orbiter, for example, ended up with engines that, by itself, it couldn't use because they had fuel only when the orbiter was connected to the external tank. The diameter of the external tank is another example of the integrated approach used in designing the entire shuttle system, according to retired NASA engineer Myron "Mike" Pessin, who spent the bulk of his career working with the external tank. Taken as a single element, there is no reason for the tank to have its 27.6-foot diameter. There was a constraint to the diameter of the solid rocket booster, however—it would have to be transported by rail from Utah to Florida, and so it was designed with a train's dimensions in mind. That diameter determined the length of the boosters, which in turn established the range of locations for the explosive bolts that connected the boosters to the external tank. Given that engineers wanted to keep the connecting bolts off of the liquid hydrogen and liquid oxygen tanks inside the external tank, they were able to establish exactly how long the structure would need to be to make that possible. Since they knew how much fuel the tank would need to hold, they could use the volume and the length to determine the needed diameter. Thus the diameter of the external tank was indirectly determined by the dimensions of the train that would be carrying the solids.

Another important part of the shuttle system design process involved computer technology that had evolved substantially since the development of NASA's earlier manned space programs. "Now we get into the hard part of, okay, now we know the requirements, how do you make this all happen?" Mattingly said.

And that all settled down certainly after Skylab, and maybe even after ASTP [the Apollo-Soyuz Test Project]. Then we started working. I remember Phil Shaffer was designated as the lead for pulling together all of our software and stuff. Because the shuttle is such a highly integrated vehicle, it has the [software] architecture that makes the system run, and then it's got all of the applications which are the heart of the vehicle. And so we were building all of this from scratch, and in Apollo we were astounded we had computers. I guess Gemini had a little computer, and then Apollo had something which, by today's standards, your wristwatch is far more powerful than what we had those days. But we were still astounded with what you could do with these things. Now we were going to build this shuttle with these computers and they're going to be its lifeblood. There won't be a lot of direct wire. Everything goes on a data bus, and this was all relatively new for most of us.

It meant learning a whole new design process, and we learned that the software was the pacing item. We blamed it on software. When we think of developing software, we think of it as coding, "if/or" statements and counting bits, but in fact the massive amount of energy went center-wide into collecting the requirements—what does it have to do, write it down, and then see if you can package it, before anybody could start worrying about building. That was an extraordinary operation. Phil drove that thing. I'm sure if Phil hadn't been there, there would have been somebody that could have done it, but I have a hard time imagining anybody that could have done it the way he did. He just had the extraordinary personality and insight. He knew all the key players from the Apollo days, and they just set out and they went to work, and they really made the program go.

In spite of all the delays that the shuttle program experienced—and we generally tended to blame that on truncated budgets, maybe some more money would have held the schedule a little better—the best I could tell, we were working as fast as that group of people [could]. It was such a massive job, and it just took so long to get everybody educated up to the same level, because it was all integrated. I don't think when we started anybody knew that it was going to be such a challenge, and so we learned to do those things and went through it. This doesn't sound like a CB [Astronaut Office] perspective, but . . . a little more than half [of the astronauts] were working the engineering side, working on the development of these things and trying to look ahead to see what was going to be required as part of getting started.

"We not only wanted to land on ten-thousand-foot runways, but we were going to be an airline," Mattingly said, explaining that since the shuttle would be a reusable aircraft with, ideally, a short turnaround time, NASA decided to turn to airline officials for help with how to do that.

So people went out and got contracts with American Airlines to teach us how to do maintenance and training, and we had people come in and start giving classes on how you give instructional courses and how we do logistics [in] the airlines. For a couple of years, we studiously tried to follow all that, and finally after a good bit it became clear that, you know, if there is anybody that's going to explain this to someone, it's going to have to be us explaining it to ourselves. That's where it evolved back into the way we had done things in the earlier programs.

Developing the systems was very much a group effort, Mattingly recalled.

I remember when we first started building the flight control schematics. Those are the most magnificent educational tools I've ever seen. I've never encountered them in any other organization. I don't know why. I used to carry around a couple of samples and give them to people and say, "This is what you really need." And they'd say, "Oh, that's all very interesting," and then nothing ever seemed to happen. But working with people to put those drawings together, and then understand what they meant and develop procedures and things from, was a massive effort. During those days the Building 4 [at Johnson Space Center] and the building behind that, where flight control teams had some other offices, the walls were just papered with these things. People would go around, and they'd walk by it and look at it, and they'd say, "That's not right." They'd draw a little red thing on it and say, "See me." And it was an evolutionary process going on continuously.

The shuttle was built with redundant systems. The idea was it should be able to suffer loss of any piece of equipment and still be able to fly safely. It was called "fail op, fail safe," meaning that one failure wouldn't affect normal operations and that a second failure could affect the way the vehicle operated but not its safety.

That generally led to a concept of four parallel strings of everything. And that was great, but now how do you manage it, and what do you do with it? Now, a schematic has all of these four strings of things, sometimes they're interconnected, and you could study those things, you'd pull those long sheets out, and you go absolutely bonkers—"Oh no. This line's hooked to that. I forgot that." Trying to figure out how this all works. So you'd go get your colored pencils out, and you'd color-code them. By now the stack of these things is building up, and I'm really getting frustrated in doing this dog-work job just before—I had to spend many, many hours for each drawing to get it sorted out before you were ready to use the drawing. So I said, "We've got to take these things and get them printed in color, right off the bat."

And so my friends in the training department said, "Well, you're probably going to have to talk to Kranz about that. He's not that enthusiastic about it." I thought, "Oh God." So I got an audience with Gene and went over and sat in his office and explained to him what we were doing in trying to get the training program started and how we were trying to get ready to do that, and I really wanted to get these things printed in color so that it would make it easier for people. I knew color printing would be a little more expensive, but it would sure save a lot of time. He said, "No. We're not going to do that." I was just overwhelmed. I said, "Gene, why?" He didn't say a word, he just turned and looked at his desk, and there on his desk, right in the corner, was this big mug filled with colored pencils. And he says, "That's how you learn." And so that was the end of the story. I don't know, I'll bet today they're still black and white. But that was Gene's method of learning, and he figured that by having to trace it out, he had learned a lot, so he felt that others would benefit from that exercise. Even if they didn't appreciate it, they would benefit.

The process of how the orbiter cockpit was designed would produce rather interesting and, in some cases, counterintuitive results, Mattingly said. He was part of a working group on controls and displays with fellow astronaut Gordon Fullerton, which made decisions about the center console.

If you sit in the orbiter, the pilot and commander are sitting side by side in the center console. It was one of the few places when, if you put on a pressure suit,

. . . you could see and touch. I mean, you can see the instrument panel. Stuff up here gets really above your head, gets really hard to see. It's in close, so it's difficult for some of us older people to focus, and you can't see a lot. You have to do it by feel, which isn't a good thing to do with important things. So the mobility was small, and this was prime real estate. We all knew it. As we went on with the program, every time someone said, "Oh, we'll just put this here [in the center console]," we'd say, "No." We'd have a big office meeting. We'd all agree that, no, that's not that important. We can put that here, we can do this. Well, after working on this thing for years, there's practically nothing that's important on the center console. We kept relegating everything to somewhere else, and it's now the place where you set your coffee when you're in the [simulator]. We protected that so hard, and poor old Gordo fought and fought for different things, and we'd think something was good, and then after we'd learn about what it really did and how it worked, we'd say, "No. You don't need that."

Then there was the question of how the Space Shuttle would fly. Each airplane flies slightly differently, or feels slightly different to a pilot flying it, and the only way to really understand exactly how a plane flies is to fly it. Further, a pilot's understanding of how airplanes fly is, to some extent, limited by the variety of airplanes he or she has flown. Those differences are rooted in the physical differences in the airplane's control systems, a factor that means something entirely different with the computer-aided fly-by-wire controls of the shuttle. "There is a military spec that publishes about flying qualities, handling qualities of airplanes," Mattingly said.

It started back in World War II, I guess, maybe even before. It tells you all of the characteristics that have to go into making a good airplane, like how many pounds of force do you put on a rudder pedal to push it. Well, even dumb pilots finally figured out that with an electric airplane this maybe isn't really relevant. Then the engineers wanted to just throw out all of the experience and say, "Hey, we'll just make it cool and you'll like it." So we went on a crusade to rewrite this document, which turned out to be one of the most interesting projects I've ever been in, because it required rethinking a lot of the things that we all took for gospel. Every airplane that a pilot flies is the Bible on how airplanes fly. Fortunately, in the office we had people who had flown a lot of different kinds of airplanes. But nevertheless, that shapes your image. And now you get into something that's totally different, and there's a tendency to want to make this new airplane fly like the one you like the most. The software guys contributed to this bad habit by saying, "Hey, it's software. You tell us what you want, we'll make it fly." I remember one time they gave us a proposal that had a little dial and you could make it a P-51 or a T-33 or a F-86 or a 747. "Just tell me what you want." We had a lot of naive ideas when we started.

While the computer for the Space Shuttle allowed many things that were groundbreaking at the time in the world of avionics, Mattingly pointed out that they were still quite primitive compared to modern standards.

I don't remember the original size of the computer, but it had a memory that was miniscule by today's standards, but it was huge compared to Apollo. By the time we finished this program, we had this horrendous debate about going to what we called double-density memory that would expand it. It was still nothing, and the only reason management did not want to change to it was for philosophic reasons. And IBM finally said, "Look, you guys said you wanted to buy off-the-shelf hardware. Let me tell you, you are the only people in the world with that version of a computer. So if you want to stay with the rest of the world, you're going to have to take this one." And fortunately, we did, and still it was miniscule. Today I think they've upgraded it several more times so that it isn't nearly the challenge. But that caused us to partition the functions in prelaunch and ascent and then get out of orbit and do some servicing things and then another load for reentry.

Don Peterson, who was selected as an astronaut in 1969 and flew one shuttle mission, said the orbiter computer systems were quite complicated.

My little desktop computer at home is about a hundred times faster and it has about a hundred times more capacity than the computers that were flying on the orbiter. They were afraid to change the computers very much because part of the flight control scheme is based on timing. If you change the computer, you change the timing, and you'd have to redo all the testing. There are thousands of hours of testing that have gone into there, and they know this thing works, and they're very loathe to make those kinds of changes. They can't change the outside of the vehicle for the same reason; that affects the aerodynamics. So they can change some things in that vehicle, and they [improved] some of it. But they're not going to make big, drastic changes to the control systems. It's just too complicated and too costly. The flight control system on the orbiter is almost an experimental design. In other words, they built the system and then they tested it and tested it and tested it. They just kept changing little bits and pieces, primarily in the software, until it all worked. But if you went back and looked at it from a theoretical point of view, that's not very pretty. You know what I mean? It's like, gee, there doesn't seem to be any consistent deep underlying theory here. It's all patchwork and it's all pieced together. And in a sense, that's true. But that's why they would be very loathe to try to make big changes to that, because putting all that stuff together took a long, long time.

Working on a project with so many systems that all had to be integrated but that were being developed simultaneously was an interesting challenge, recalled Mattingly. "Within the office, we were all trying to stay in touch with all these things going on in each of these areas to keep them somewhat in sync from the cockpit perspective. So that gave us a lot of insight into all of these tasks that people were doing," he said.

We even found, for instance, that as part of this development program, people working with thermal protections systems, the structure guys found that they were discovering limitations that were going to be imposed on the vehicle downstream that we weren't thinking about—if you fly in the wrong regimes, you will get yourself into thermal problems. Yet nothing in our flight control work or displays was considering that. We had never encountered anything like that before. So the guys, by working all these different shops, were picking up these little tidbits and we were trying to find ways to look ahead.

Another major change, Mattingly said, was developing and testing the flight control software for the shuttle. "We learned quickly that the man-machine interface is the most labor intensive part of building all this software," he said, explaining that the code dedicated to computer control of the vehicle made up less of the software—and less of the time it took to develop it—than the code related to the interface that would allow the astronauts to control use of that software to control the vehicle. In addition, he said, a conflict arose because of the computer use needed to develop and test that software. To the engineers who were using those computers to design the vehicle, the time the astronauts spent testing and practicing with the flight control software seemed like "video games."

We ended up building a team of people: Joe Gamble, who was working the aerodynamics; Jon Harpold, doing guidance; and Ernie Smith, who was the flight control guy. They all worked in E&D [Engineering and Development]. We all got to going around together in a little team, and we would all go to the simulators together, and we would all study things. We built a simulator from Apollo hardware that was called . . . ITS, the Interim Test Station. We had a couple of people—Roger Burke and Al Ragsdale were two sim engineers that had worked on the CMS [Command Module Simulator] and the LMS [Lunar Module Simulator]. They were very innovative, and they took these things before we had the Shuttle Mission Simulator that was back in the early part of the design and went to the junkyard and found airplane parts and built an instrument panel out of spare parts and had a regular chair that you sat in and had different control devices that we had borrowed and stolen from places. These folks were so innovative; they could hook it all up.

"They took the initial aerodynamic data books and put them in a file so we could build something that would try to fly," Mattingly said.

We even took the lunar landing scene television. In the Lunar Module Simulator they had a camera that was driven by the model of the motion and it would fly down over the lunar surface, and so you can see this thing, and that was portrayed in the LMS as what you'd train to. So they adapted that to a runway. We tried to build a little visual so we could have some clues to this thing, put in a little rinky-dink CRT [cathode-ray tube] so we could play with building displays. And we got no support from anybody. I mean, this wasn't space stuff. And it is probably one of those things I was most proud of, because we were able to get this thing into someplace where we could actually tinker with how we're going to fly the vehicle and what we're going to do and what the aerodynamics mean. It was only possible because we had these two simulator guys who were wizards at playing with software and this team from E&D who joined us.

We ended up realizing that we had built an electric airplane that had essentially only one operating flight control system. So we said, "Well, what if we're wrong? No one has ever flown a Mach 20 airplane. This whole flight envelope is something that nobody's ever had the opportunity to experience. So what do you suppose our tolerance is to this?" Because wind tunnel models for the ascent vehicles, they fit in your hand, because the tunnels that were able to handle these things were small. The wind tunnel models for the orbiter were larger, but they're still not all that big, and going through this tremendously wide flight regime where the air density is going from nothing to everything, and it's just high speeds to low speeds, I said, "What's the chance of getting all that right?" And yet as we played in these simulators, . . . we proved to ourselves that, boy, if you're off on that estimate of the aerodynamics, you can often play with the software to make it right, but if the real aerodynamics and the software you have don't match, it's a real mess. I know I worried a lot about that.

So we came up with a concept that we would have some tolerances on the aerodynamics, and we would try to make sure that the flight control system could handle these kind of uncertainties in aerodynamics. We did something which is not typically done—we decided to optimize the flight control performance to be tolerant on uncertainties rather than the best flight control system they could build. The whole idea was, after we've flown and we have some experience and we know what the real world is, now we can come back and make it better, but the first job is to make ours as tolerant as possible to the things we don't know.

While Mattingly was working with the computer models of the flight dynamics of the shuttle, astronaut Hank Hartsfield was on the other side of that research, working with the wind tunnel models and encountering the same concerns about the scalability of the data coming out of those tests.

As I recall, the shuttle program had over twenty-two thousand hours of wind tunnel time to try to figure out what it flies like. Because the decision had been made, there are no test flights. We were going to fly it manned the first flight, and an orbital flight at that, which demanded that, the best you can, [we] understand this. Well, hypersonic aerodynamics is difficult to understand, the uncertainty on the aerodynamic parameters that you get out of the tunnel are big. The things that we were looking at in the simulations were if these uncertainties in the different aerodynamic parameters stack in a certain way, the vehicle could be unstable.

What we were looking for, for those combinations, statistically were possible, but hopefully not very probable they'd happen, but if they did, that was the kind of things we had to plan for. It's just an uncertain world. You can't predict, because in the wind tunnel, you have to put in scaling factors. If you're doing wind tunnel things off a small model, it doesn't really scale to the big model perfectly, and you have to make assumptions when you do that. The scaling ratios have a big factor, a big effect on what the real numbers are. So if you could fly a full-scale orbiter in the wind tunnel and it would go Mach 15 or something, it would be great, but you can't do that. You have a little-bitty model, and it's a

6. Space Shuttle vehicle testing in the fourteen-foot Transonic Wind Tunnel at NASA's Ames Research Center. Courtesy NASA.

shock tunnel or something. You'd get a few seconds of runtime at the right Mach numbers and then try to capture the data off of that.

Astronaut Don Peterson was involved in studying the redundancy of systems on the orbiter, and particularly the flight control computers. In the report he pointed out that failure rates on some of the avionics could be high. On Apollo and earlier vehicles, NASA built "ultra-reliability components," components that were overdesigned and tested to make failures less likely.

Failures on Apollo, for that reason, were pretty rare. But that's very expensive. That's a very difficult thing to do. I was told that after the lunar program ended, MIT had two of the lunar module computers left over, spares. So they just turned them on and programmed them to run cyclically through all the programs. I think they ran one of those computers for, like, fifteen years, and it never failed. It just kept running, and finally they turned it off. They just said, "It's not ever going to fail." That's the way that equipment was built. But that makes it very expensive. So when they built the Shuttle, they said, "We can't do that. So what we're going to do is, instead of ultrareliability components, we're going to rely on something called redundancy." They were going to have four computers, and they were going to have three TACANs, *and they were going to have four of this and two of that and so on. That way, you could tolerate failures. But as a result of that, the failure rate on some of that equipment was fairly high, compared to Apollo.*

They also made the multiple units interdependent. "On a typical automobile you have five tires, but that's not five levels of redundancy because you need four of them," Peterson explained.

So you can really only tolerate one failure. You can have one tire go bad and you can take care of that. But we got into that same situation on the shuttle because of the way they did the software. The shuttle, when it's flying, the computers all compare answers with one another, and then they vote among themselves to see if anybody's gone nuts. If a computer has gone bad, the other computers can override its output so that it isn't commanding anything. But to make that scheme work, you have to have at least three computers working. Otherwise, you can't vote. You could have [two systems voting], but if they vote against each other, you don't know which one's the bad one.

The decision was made to put five of the computers on the orbiter, with four of them active in the primary system, with the idea that this would create a system that could tolerate three failures. However, Peterson said, this produced much higher failure rates than expected. While the system provided a high amount of redundancy in theory, the reality was that because of the way it was designed, the system actually could tolerate only one failure safely. The four primary computers were not truly redundant for each other; only the spare provided redundancy. If one computer failed, the spare would take its place. After that, however, further failures would endanger the cooperative "voting logic" between the computers that verified the accuracy of their results.

But the complexity of the way the thing was put together kind of defeated the simplistic redundancy scheme that they had. It'd be like driving a car that had two engines or three engines, and any one of them would work. Well, that way you could fail two engines and you'd still drive right along. But if it takes two engines to power the vehicle, then you don't have that, and if it takes three en-

gines to power the vehicle, you don't have any redundancy at all. It gets to be a game then as to how you trade all this off. When I looked at all that and we put the study together, we said, "You know, you're going to have some failures that are going to really bother you because you're going to lose components." For example, you're on orbit and you've got four computers and one of them fails. Well, now you've got three computers left in the primary set. But do you stay on orbit? Because if you suffer one more failure, your voting algorithm no longer works. Now you're down then into coming home on a single computer and trusting it. And nobody wanted to do that.

So they said, "Gee, I've got four computers. I can only tolerate one failure, and then I've got to come home." We had four of some of the other components, and it was kind of the same sort of thing. If one of them fails, we are no longer failure tolerant. We've lost the capability to compare results and vote, and so we don't want to stay on orbit that way. So now, all of a sudden, the fact that you've got four of them causes more aborts because the more things you have, the more likely you are to have one fail. You'd get more failures and more aborts with four computers than if you'd gone with some other plan. That was pretty controversial for a while. We predicted—and there were some people that were really upset about that—we predicted a couple of ground aborts due to computer failures. Essentially we'd get chewed out for saying that, but in the first thirteen flights, we hit it right on the money. We had two ground aborts in thirteen flights.

When the shuttle was built, the air force was also using redundancy systems, Peterson recalled. Then the air force built what it called confederated systems, in which each component was independent. "They cooperated with each other, but they shipped data to each other, but they weren't really closely tied together," Peterson explained.

The shuttle was tightly integrated. It runs on a very rigid timing scheme. The computers on the shuttle actually compare results about a little more than three hundred times a second. So it's all tightly tied together. Well, when they decided to build the [International] Space Station, NASA *said, "We're not doing this integrated stuff anymore. Boy, that was a real pain. We're going to use a confederated system." The air force, on their latest fighter, said, "This confederated stuff doesn't work worth a damn. We're going to build a tightly integrated [system]." So they both went along for ten years or twelve years, and then they flip-flopped. The military's going the way* NASA *originally went, and* NASA*'s now going the way* *the military went originally. I think the answer is, there is no magic answer to all that. Probably one concept is maybe not that much better than the other. It's how you implement it and how much money you spend and how much to test. What do they say? The devil's in the details. I think that's right with all this stuff.*

Mattingly recalled excellent cooperation between the engineering staff working on the shuttle and the Astronaut Office. "I seldom have seen that integration of the people that were going to fly it with the designers and people who were doing the theoretical work and the operators from the ground," Mattingly said.

All of that stuff was converged in parallel, and I think that's one of the reasons that the shuttle is such a magnificent flying machine. It does all the magic that we set out to do. I'm ignoring the cost because the shuttle, in my recollection, by the time it was sold to Congress, it was probably different than what the people in the trenches remember, but we had to do all these technical things, and it was a matter of faith that if you build it, it will be cheap. I mean, it was just simple. If you could reuse it, it saves money, and so you've got to make it reusable. If you fly a lot, that will be good, and we're going to fly this thing for $5.95, and we're going to fly it once a week and that's how we're going to do this. And none of us were ever told to go build a vehicle that we could afford to own. And had we been told that, I doubt if we would have been able to do it. I think the job was so complex, you had to build one that flies in order to learn the lessons that say, "Now I know what's important and what isn't." I just think it would have been asking too much, but that's just personal opinion, but it's from having struggled through ten years of this development program. It was an extraordinary experience to do that.

The role of the Astronaut Office during the development of the Space Shuttle was quite different from what Mattingly experienced during the Apollo program. "Our involvement was far more extensive and pervasive, and a heck of a lot more fun," he said.

I mean, this was really cool stuff. There was a problem every day, and you got to learn about all of these little things that were interesting. I spent a lot of time trying to understand the stress loads and the thermal characteristics on the TPS [thermal protection system], and how do you get it to stay on, and all of those things were things that came through the office as experiences that really were just extraordinary opportunities to go see that. As we moved down the stream and we got into some of these development programs and started turning out hardware, we started splitting people up to go follow different components of hardware, whether it be the engines or the SRBs *or the orbiter.*

The decision to have the orbiter be an unpowered glider rather than a jet during its return to Earth and the various ramifications of that decision were also among the things that had to be considered during development. "Somewhere earlier in this development stage, we went through a series of activities where the first orbiter was going to have air-breathing engines, and it had some solid rockets that were on the back that were for aborts," Mattingly said.

Right off the pad you could fire these two big rockets, and they would take you off in a big loop so you could come back and land. We had these air-breathing engines that were going to—after you come down through the atmosphere, you open the door and these engines come out, and you light them and you come around and land. They had enough gas for one go-around. The other thing we had was the big solids were to have thrust terminations and ports that blew out at the front end so you could terminate thrust on them if you needed to in an emergency. Every one of those devices was something which had a higher probability of killing you by its presence than it would ever have in saving you. I'll put that ejection seat in the same boat. Everybody was willing to get rid of the air-breathing engines. They were really, really not a very bright idea. And we got rid of the thrust termination and we got rid of the abort solid rockets. My guess is John Young was probably the most active stimulus in pushing those issues, and that was one of those cases where the flight crew perspective and the engineering perspectives converged. We all wanted to get rid of these things, and yet we retained the ejection seats for reasons which I will never understand. If anyone knew what the useful envelope of those ejection seats was and the price we paid to have them. . . . But it had become a cause: "You will protect these kids by giving them an ejection seat." So we had one, not that anybody wanted to ever use it, but it was there.

Astronaut Bonnie Dunbar was still an undergraduate student at the University of Washington during the early portions of shuttle's development, and she worked with the school's dean of ceramics engineering, who had received a grant to work on the tiles for the shuttle's thermal protection sys-

7. A worker removes a tile as part of routine maintenance activities on the orbiter fleet. Courtesy NASA.

tem. NASA's earlier manned spacecraft had used ablative heat shields, which absorbed heat by burning up, protecting the rest of the vehicle. Such a system was simple and effective, but for the new, reusable Space Shuttle, NASA wanted a reusable heat shield, one that could protect the vehicle without itself being destroyed. The solution that was settled upon involved a vast collection of tiles and "blankets" covering the underside of the orbiter and other areas of the vehicle that would be exposed to extreme temperatures.

"First of all, tiles are a ceramic material, so by definition they're brittle," Dunbar said.

But the reason they have an advantage over metals is that they don't expand ten times over their thermal exposure range. It's called the coefficient of thermal expansion. Also, they are an insulator; they don't conduct heat. We looked at metals, or what they call refractory metal skins, and there are two disadvantages. You still have to insulate behind them, because metals conduct heat. The other is that when you go from room temperature, let's say seventy-five degrees Fahrenheit, to twenty-three hundred [degrees], you have a large growth. It's like your cookie pans, I guess, in the oven. So the airframe would distort. The ceramic materials [have] very small thermal coefficients of expansion, ten to the negative sixth, so you're not going to see a lot of deformation. Also you could, on a very low density tile, expose the surface to twenty-three hundred degrees Fahrenheit, and the backface, three inches deep, would not see even close to that, less than a couple hundred degrees, till after you're on the ground. It's a very slow coefficient of thermal expansion and heat transfer. So ceramics had a definite advantage. We knew that from the work we'd done in the sixties, and in fact, ceramics were already being used as the heat shields on nose cones for missiles and so forth. So the next big challenge was to put them in a low-density, lightweight form that could be applied to the outside of a vehicle. Apollo vehicles, Gemini, Mercury, were all covered by ablators, which meant that they burned up on the reentry to the Earth's atmosphere and could not be reused. The tiles were meant to be reusable. They didn't deform. They didn't change their chemistry. We had to, though, shape them so that they were the shape of an airplane, so we had all the aerodynamic features there. So we sort of did a little reverse engineering, in that we said, "Okay, here's what the shuttle looks like; got to maintain that shape. Here's how hot it gets from the nose to the tail. Most of the heat's at the nose, on the nose cone, and the leading edges of the wings. We want to make sure the aluminum substructure doesn't get over 350 degrees Fahrenheit; that's when it starts to change shape. So how thick does the tile have to be?" So we used all those limits and constraints, then you'd use the computer . . . to calculate how thick each tile had to be. Then we started looking at, well, okay, how big should each tile have to be? Could I just put large sheets of tile on there?

Well, we started looking at what the structure does during launch, and now we're getting to something called vibroacoustics. There's a lot of force pressure on the vehicle, a lot of noise, if you will, generated into the structure, and it vibrates. We calculated that if we put a foot-by-foot piece of tile on there, the vibration would actually break it up into six-by-six-inch pieces. We said, "Well, we'll design it six by six." So you'll see most tiles are six by six. Now how close do you put them? We thought, well, you can't get them too close, because during that vibration they'll beat each other to death, because they're covered with a glaze. You've got silicon dioxide fibers that are made into very low mass tiles, nine pounds per cubic feet, or twenty-two pounds, and to ensure they don't erode in the airstream when you reenter, they're covered with a ceramic glaze. So that's also brittle, so you can't get them too close or they'll break the glaze. You can't get them too far apart or, during reentry, the plasma flow will penetrate down in those gaps and could melt the aluminum. So that's called gap or plasma intrusion. So that then constrained what we called the gap. Then from tile to tile, how high one was compared to the next one, we called step. That became important because if you had too large a step towards the leading edge of the wing, that would disturb the boundary layer, and you would go up the plasma, and instead of having smooth layers, it would start to transition to turbulent, from laminar to turbulent, and turbulent results in higher heating. So that controlled the step. So gap and step were very important to that as well.

"Those were all challenges," Dunbar said. "We depended on advances in computerized machining capabilities, wind tunnel work with models to help us determine the requirements on step, the manufacturing, just everything. Firing a tile, a certain temperature and time was important to maintaining its geometry. . . . It's, I think, a real tribute to the program that if you look at follow-on programs, even in NASA but also in Japan or in Europe or even the Russians, who built the Buran [Soviet shuttle], you'll find that the system on the surface is very similar to the shuttle tile system. It was a good solution."

Dunbar said that working on the shuttle during that early time was an exciting opportunity.

This was the next-generation vehicle. Not only was it next generation, it was . . . "transformational" is the word we use now. If you think about it, everything to that point was one use only. Couldn't bring any mass back. We sent a lot of things into orbit that we had to test and leave there, and it became a shooting star, coming back to Earth. So this transformed our ability to do research. It's why we have a space station now. We not only learned from Skylab, but we flew [on] Spacelab countless research projects that we could bring back to Earth, get the results out, diagnose problems with equipment. I think it saved the government billions of dollars, because we didn't throw it away each time. So it was exciting, and we knew what it could do. New technology. It was leading edge on not only the thermal protection systems, but it was the first fully fly-by-wire vehicle, in terms of the computers and the flight control system. The main engines were also a pathfinder as well, and so it was exciting, even if it delayed till '81. If you think about it, we baselined it to the contractor, to Rockwell, in 1972, I believe. So nine years later we have a vehicle, a reusable vehicle, flying.

Astronaut Terry Hart was the Astronaut Office's representative in the development of the Space Shuttle main engines.

Since I had a technical background, mostly mechanical engineering, John Young had asked me to follow the main engine development. This was a couple of years before STS-1. *In fact, it was ironic that we showed up [as* NASA *astronauts] in '78, and everyone said we're one year away from the first shuttle launch, and two years later, we were still one year away from the first shuttle launch, and it was really because of two main areas of technical difficulty. The main engine development was somewhat problematic, with some turbo pump failures that they'd had on the test stand, and the tiles. We had difficulty with the tiles being bonded on properly and staying on. But the main engine was one that John Young wanted me to follow for him, and so I spent a lot of time going back and forth to [Marshall Space Flight Center in] Huntsville [Alabama] and to* NSTL, *the National Space Technology Laboratories, in Bay St. Louis [Mississippi, currently called the* NASA *John C. Stennis Space Center], where* NASA *tested the engines. And Huntsville, of course, was where the program office was for the main engines. And that was very exciting. I mean, I was like a kid in a candy store, in the sense that a mechanical engineer being able to kibitz in this technology, with the tremendous power of the fuel pumps and the oxidizer pumps, and the whole engine design, I thought, was just phenomenal. The hard part of that job was when we had failures on the test stand, which were, unfortunately, too frequent. I'd get the pleasure of standing up in front of John Young and the rest of the astronauts on Monday morning to explain what happened. And, of course, everyone was always very disappointed, because we knew this was setting back the first launch and it was a jeopardy to the whole program. But we got through that, and the engines have done extremely well all through the program here, where it was always thought to be the weak link in the design.*

Astronaut Don Lind was involved in the early planning and development of the remote manipulator system, the shuttle's robot arm.

I guess the first significant assignment I had [for the shuttle] was in developing the control system for the remote manipulator system, the RMS. *In the hinge line of the cargo bay doors, there is an arm that's articulated pretty much like the human arm. It's about as long as two telephone poles, and it's designed for deploying and retrieving satellites. Again, somebody had to worry about the operational considerations of that arm. It was built by the Canadians with the agreement through the [U.S.] State Department, and I was assigned to work on that. So I made a lot of trips into Canada to work with those people. The peo-*

ple who were actually building the hardware were very, very compatible, very easy to work with, and we had a very nice working relationship.

Lind contributed to the development of the three different coordinate systems that were going to be built into the arm's software.

One coordinate system, obviously, applied when you're looking out of the window into the cargo bay, and so you want to work in that coordinate system. If you wanted the arm to move away from you, you pushed the hand controller away from you. Also, if you're trying to grasp a satellite up over your head and you're looking with the TV camera down the fingers at the end of the arm, which is called the end effector, and you want to move straight along the direction the fingers are pointing, you don't want to have to try to figure out which way you should go, so you shift to a totally different coordinate system. So if you're looking in the TV picture with the camera that's mounted right above the end effector, you want to push the hand controller straight forward. You want it to move straight forward in the television picture.

Lind also helped answer the question of how the hand controllers were to be configured.

We wanted hand controllers where the translation [movement] motion would be done by one hand controller, which we decided would be the left hand, and the rotational motion controlled by a hand controller which would be handled with the right hand. We decided, as a joint decision, that the hand controller for translation should be a square knob.

Then I said, "Now, remember you're floating. You're floating, so you've got to hang on to something while you're translating, and you don't want your bobbing around to affect the hand controller. So you need to put a square bracket around it so you can hold on to the bracket with your little finger and can use the hand controller." "Oh yeah, we hadn't thought of that. Well, how big do you want it to be?" We actually measured my hand and designed the controller and bracket to the physical dimensions of my hand. Obviously, when you make a decision like that, then you have five other astronauts check it out, and they say, "Yeah, that was a really good decision." I didn't want the hand controller for the right hand to be mounted square on the bulkhead, because the relaxed position of your arm is not at a square angle; it's drooping down to the side. And I wanted that position to be the no-rotation position. We set up a simulation, and I stood up there, and they measured the angle of my arm and then built a bracket to mount that hand controller just exactly the way my arm relaxed. And again, we had several other astronauts check it, and they said yes, that was a fine thing. So the hand controllers were literally fine-tuned to my design.

Other people were worrying about the software, how to implement these coordinate systems. Other people were doing all the very sophisticated engineering. But the human factor was my responsibility, and basically it was a very pleasant experience to work with the Canadians, with one exception. The arm has two joints: like the elbow, and like the shoulder; one degree of freedom in the elbow, two in the shoulder, and three degrees of freedom in the wrist, so there are three literal components to the wrist junction. They had mounted the camera on the middle one. As you maneuver in certain ways, the wrist has to compensate for the rotations of the other joints, and every once in a while the TV picture would simply rotate. Not that anything had actually rotated, but the wrist was compensating. I said, "That's unacceptable." They said, "No, no, no, no, it has to be there. That's the cheapest place to put it." The engineers were all in agreement that this was a mistake, because you could lose a satellite when suddenly the picture rotates and nothing really has happened. But the management people said, "This meets our letter of intent with the State Department. We're not going to change it." So in one meeting I had to be very unpleasant. I said, "Now, gentlemen, if we ever lose a satellite because of this unnatural rotation, I will personally hold a press conference and say that you had been warned, and it's the Canadians' fault." They looked at me like, "Ooh, you're nasty." At the next meeting, they said, "Well, we'll change it, and it doesn't cost as much as we thought in the first place." Usually you could get good cooperation, but occasionally, particularly with people up in the bureaucratic levels, you had to be a little bit pushy. I try not to be pushy, but that's one time I did.

Astronaut George "Pinky" Nelson was involved in the development of the Extravehicular Mobility Unit (EMU), the spacesuit used for conducting activities outside of the spacecraft. "The suit was one of the long poles in getting the shuttle ready to fly," he said.

The folks in Houston who were in charge of it, [Walter] Guy and his group, were really working hard, and it was a difficult task to get it pulled together. The suit actually blew up shortly before STS-1. I was home working in my garden. I was playing hooky one afternoon, and I got a call from George Abbey.

He said, "Where the hell are you?" "Well, I'm home working in the garden." He said, "Okay. Get in here. We just had an accident with the spacesuit." They were doing some testing in one of the vacuum chambers in Building 7, and they had the suit unmanned, pressurized, in the vacuum chamber. They were going to do some tests and they were going through the procedures of donning the suit and flipping all the switches in the right order and going through the checklist. There's a point when you get in the suit that you move a valve. There's a slider valve on the front of the suit, and you move this slider valve over, and what it does is it pushes a lever inside a regulator and opens up a line that brings the high-pressure emergency [oxygen] tanks on line. You do that just before you go outside. You don't need them when you're in the cabin, because you can always repressurize the airlock. When you're going to go outside, you need these high-pressure tanks. They're two little stainless steel tanks about six inches in diameter, maybe seven. And it turned out that when this tech did that, he threw that switch and the suit basically blew up. I mean not just pneumatically, but burst into flames [and] got severely burned. It was pure oxygen in there. The backpack is made basically out of a big block of aluminum, and aluminum is flammable in pure oxygen. So this thing just went "whooff," went up in smoke.

So then I was put on the Investigation Board for that, and spent I don't know how long, a couple months at least, just focusing on what had caused this and could we identify it and fix it and get it ready so that it wasn't the long pole for flying STS-1. So I learned even more about the design and manufacturing and materials and all of that in the suit during that process. It was fascinating. And the NASA system for handling that kind of an incident really is very good. We've seen it with the big accidents we've had. They really can get to the bottom of a problem very well.

After that, Nelson said, there weren't any major problems in the development of the suit. "There were lots of little stuff. The displays and controls on the suit are a challenge because, one, you have to see them from inside the suit, looking down, so a lot of these old guys in the office who were, you know, the stage I am in my life now, where I have to wear reading glasses, couldn't read the displays because they were close to your face. So we worked on lenses and all kinds of ways to make the displays legible to people with old eyes."

For all the capabilities built into the vehicle, one of the notorious disappointments of the Space Shuttle program is that launch costs ended up being much higher than promised. The original appeal of the shuttle was that its reusability would bring launch costs down dramatically, but those dreams were never fully realized. Explained Don Peterson,

The shuttles, unfortunately, are pretty difficult to work on. When the military builds an airplane, it tries to make everything in the airplane designed so that you can remove and replace parts quickly and easily. The shuttle is much more difficult to get to some of the stuff. There're not big [easily opened] panels on it. You can't release a few latches and open a big panel on the side of the orbiter. You literally have to take it apart to get into it. You can go in through the inside, through the bay, and get to some of that stuff, but even then you're removing parts that aren't designed [for that]. It's not like opening doors and looking inside. The military builds a lot of their stuff to be easy to work on, and they really didn't build the shuttle that way. So the shuttle is more expensive to operate. For example, the little jet engines, there's, like, thirty-eight of them, I think, on the orbiter that control attitude when it's on orbit. If one of those engines fails, you can't just unscrew some things and take it out. You have to cut it out with a torch, and you have to weld the new one back in, because they didn't build it to be removed. The heat shield is [24,300] little individual tiles, and they're all different shapes and different thicknesses, and so every tile is like a little individual item. When the shuttle comes back, they have to inspect visually, and with a pull device, every single tile. If any of them don't pass, you've got to cut that one out and clean off the glue and go get the new one and put it all back. Those are very high maintenance items. So the shuttle really wasn't built to be easy to maintain, and that's because NASA *has always had, as [former Johnson Space Center director] Gerry Griffin used to say, a standing army at the Cape that did all that, and nobody really worried about it. If you needed something done, you just called and they sent over four or five guys and they fixed it. But that's expensive.*

The shuttle was designed to fly, I think it was fifty flights a year, and they were going to have five shuttles to do that. So each shuttle would fly ten times in a year. Well, right now the whole fleet's only flying about eight times a year. Well, you're trying to amortize the cost of the whole program over eight flights. It's like we've got all this capability to repair and replace and analyze and monitor things, and we're not using a whole lot of it. If you were flying fifty times a year, the cost per flight would go way down because you wouldn't add that much to the facilities and the maintenance costs. The facilities costs don't change much if you never flew. You've still got to have all the facilities, and you've got to pay for all that. You have to keep this whole group of specialists on, technicians and people, to do the work. With eight flights a year, some of those guys may only get used twice a year, but you've got to pay them and you've still got to have them there. If you were flying a lot more, the cost per flight would go way down.

George Mueller, the NASA head of human spaceflight who launched the Space Shuttle program, explained that there were several factors that drove the operational cost of the shuttle up, including many decisions, like the use of solid rocket boosters, that reduced development costs at the outset and presented Congress a lower buy-in budget request to build the vehicle but that resulted in higher operational costs once the shuttle started flying. However, he said, the ultimate problem with the shuttle was that it ended up being designed to use far more people to process it than were absolutely necessary. "If you really want to know why the shuttle failed, it's because they designed it to use all the people from Saturn and Apollo, to keep them employed."

Countless technical problems had to be overcome, and ultimately the shuttle's greatest limitation was that it was designed to be too nice.

Former JSC director Chris Kraft, however, still speaks highly of the shuttle. "It's the safest spacecraft we ever built." Kraft noted that while shuttle crews have been lost because of problems stemming with the solid rocket boosters and the external tank, the orbiter itself has not been responsible for any fatal accidents. "The orbiter itself is flawless, since we've been flying. Absolutely flawless." Rather than retiring the shuttle, Kraft argued, NASA should have continued to make it better and continued to fly it, adding that many ideas for improving the orbiter were never implemented. "That's what we should still be doing. We still ought to be improving. We could improve the hell out of it. We could improve the hell out of the thermal protection system, we could improve the control systems, get rid of the APUs [auxiliary power units]. All of that has been designed and is ready to be built. You don't have to stop and redesign it, it's done."

3. TFNG

By 1976 NASA's astronaut corps had seen a large number of departures. Many of the early astronauts who had joined the agency as pioneers of spaceflight or as part of the race for the moon felt like they had accomplished what they had come to do. The last Saturn to fly launched in 1975, the next opportunity to fly was still years away, and some in the corps decided they had no desire to wait.

Only one of the Original Seven astronauts, Deke Slayton, remained in the agency, as did only one member of the second group, John Young. Two members each remained of the third and fourth groups (although only one of those four astronauts would get the opportunity to fly on the shuttle). The fifth group was better represented—eight of the Original Nineteen were still at NASA—and the majority of the sixth group and all of the seventh were still at the agency, having arrived in the corps too late to be assigned Apollo flights.

With the number of astronauts dwindling, the ambitious plans for the shuttle program required new blood. So in 1976 NASA announced for the first time in a decade that it would be accepting applications for a new class of astronauts, to support the Space Shuttle program.

Astronaut Fred Gregory saw the ad for Space Shuttle astronauts on television. "I was a *Star Trek* freak, and the communications officer, Lieutenant Uhura, Nichelle Nichols, showed up on TV in a blue flight suit," Gregory said. "As I recall, there was a 747 in NASA colors behind her; you could hear it. But she pointed at me and she said, 'I want you to join the astronaut program.' So, shoot, if Lieutenant Uhura looks at me and tells me that, that got me thinking about it."

Steven Hawley saw the NASA announcement on a job openings bulletin board while in graduate school at the University of California.

I remember there was this letterhead that said NASA on it, and I thought, "Wow, that'd be interesting." I looked at it, and it said they were looking for astronauts.

I had no idea how they'd go about hiring astronauts, and here's an announcement saying, hey, you want to be an astronaut, here are the qualifications. You have to be between five foot and six foot four, and you have to have good eyesight, and you have to have a college degree, and graduate school counts as experience. You need three years of experience, and I'm thinking, "Well, I'm qualified." I've also told kids that so were twenty million other guys.

Hawley recalled that this was the first time he thought that becoming an astronaut might really be possible for him, because of changes in the selection criteria. "I probably dropped everything I was doing at that moment and set about filling out this application to become an astronaut. I didn't realize till years later that it's actually the same application you fill out to be any government employee, SF-171. You fill it out and send it in. I even remember sending it by, I think, return receipt request so that I could make sure that this thing got into the hands of the proper people at NASA."

Realistically, Hawley said, he didn't think he would be selected. He realized the pool contained many well-qualified applicants. But even with what he believed were slim odds, he applied anyway. "Why in the world would they pick me?" he said.

I still think perhaps they didn't mean to, and one day they'll come and tap me on the shoulder and say, "Excuse me. You've got this guy's desk, coincidentally named Steve Hawley, and he's the one we meant." I've told kids this, too, that the reason I applied, as much as anything, was because I knew that if I applied and didn't get picked, and then I watched shuttles launch with people on them and building space stations and putting up telescopes in space, I could live with that, if NASA said, "Well, thanks, but you're not what we're looking for." But to not apply, to not try, and then wonder your whole life, could you have done it if you had tried, I didn't think I could take that. So it was okay if they said no, but I didn't want to go through the rest of my life wondering, had I only tried, would I be able to do it?

Before 1978 NASA had selected five groups of pilots and two groups of scientist-astronauts. The eighth group would be the first mixed class, including both pilots and a new designation, mission specialists.

The new designation was of particular interest to Mike Mullane, who at the time was a flight-test weapon system operator for the air force. "NASA announced they were selecting mission specialist astronauts, and this was the new thing, because now you didn't have to be a pilot to apply to be an astronaut. So this dream of perhaps being an astronaut was now back open to me. In fact, I remember that night that they announced it. This was big news at Edwards, because virtually everybody at Edwards Air Force Base wanted to apply to be an astronaut."

The new class would be the largest group of astronauts yet. More than eight thousand applications were received. In 1978 NASA announced the first class of shuttle astronauts, dubbed TFNG, an acronym given multiple meanings, most politely, "thirty-five new guys."

Among the new class were, of course, test pilots from the navy and the air force, many of whom knew each other and had trained and served together. Rick Hauck was on his second cruise as a navy pilot on the USS *Enterprise* when the announcement came out. "There was a flyer from NASA saying they were looking for applicants for the astronaut program to fly the shuttle and, in fact, four of us on the *Enterprise* wound up in my astronaut class: myself, Hoot Gibson, Dale Gardner, and John Creighton. Three of the fifteen pilots were from that air wing. Dale Gardner was a mission specialist. Which is really kind of interesting, three of fifteen. What's that? Twenty percent came from that ship."

Hauck didn't grow up with an interest in space, and as a child there had been no space program for him to aspire to. "The word *Apollo* didn't even exist in terms of spaceflight when I was thinking about becoming a naval aviator," said Hauck, who was a junior in college when Alan Shepard made his first spaceflight in 1961. "Even before I became an aviator, while I was at [The U.S. Naval Test Pilot School in] Monterey, I had read that NASA was recruiting scientists to become astronauts, and I wrote a letter to NASA saying, 'I'm in graduate school. You could tailor my education however you saw fit to optimize my benefit to the program, and I'd be very interested in becoming an astronaut.' I got a letter back saying, 'Thank you very much for your interest. Don't call us. We'll call you.' That was in early '65, I think, so it was twelve years later that I was accepted into the astronaut program."

Sally Ride, the United States' first female astronaut to fly in space, saw the ad for a new class of astronauts in the Stanford University newspaper, placed there by the Center for Research on Women at Stanford. "The ad

8. Astronauts training to experience weightlessness on board the NASA KC-135. Courtesy NASA.

made it clear that NASA was looking for scientists and engineers, and it also made it clear that they were going to accept women into the astronaut corps. They wanted applications from women, which is presumably the reason the Center for Research on Women was contacted and the reason that they offered to place the ad in the Stanford student newspaper."

Another member of the eighth class, air force pilot Dick Covey, got to NASA by studying and following a career path similar to those of the early astronauts. "As I looked at what it looked like those original astronauts had done . . . that became a path for me to follow," Covey said. He majored in astronautical engineering and participated in a cooperative master's program between the Air Force Academy and Purdue University. According to Covey, fifteen of the selected Thirty-Five New Guys participated in the program at the same time as he.

We gave up our graduation vacation time. All my [other] classmates got two months to go off and party and tour the world and do whatever and then go to their flight training, while we all went immediately, right after graduation to Purdue and started school again. But in January following graduation in June, we all had our master's degree in aeronautics and astronautics, and those of us that were going to flight training already had our flight-training date, and we went immediately to flight training. So, for someone that wanted to be an astronaut, being able to go through the Air Force Academy, major in astronautical engineering, and get a master's degree from Purdue in aeronautics and astronautics within seven months and then go immediately to flight training was an extraordinary opportunity. I often wonder, if I had not done that, whether I would have ever become an astronaut. . . . One of the reasons Purdue has so many astronauts is there's all these Air Force Academy guys who went through that program over time, and it added to their numbers then.

When the announcement was made, Covey applied through the air force. The air force had decided, as the other services did, that it would have its own selection of those it would nominate to NASA, and Covey was selected as one of the air force's applicants.

Hauck and Dan Brandenstein were test pilot school classmates and squadron mates six years prior to their selection to the corps. Hauck said the two talked a lot back then about whether or not they would apply to the astronaut program.

Part of the preinterview process was the folks in Houston took each folder. Some of the people were rejected immediately. Some, they said, "Well, let's find out more about this person." They make a lot of phone calls. "Hey, do you know Rick?" or, "Do you know Dan? What do you think about him?" So I got a call one day in my office at Whidbey Island, Washington, and it was John Young. And John said, "I'm on the selection board for this astronaut program." He didn't say anything about knowing that I was applying. He said, "Dan Brandenstein, he's in your squadron there. What do you think about him?" And I told him, I said, "I think he's a great guy. He'd be a super astronaut." He said, "Okay, thank you very much." And I said, "Excuse me, but I'm applying also." He said, "I know. I know. Thank you very much."

Covey said that his selection as one of the air force's candidates for the new class of astronauts was the first of a series of milestones that made the possibility of achieving his goal seem a little more real. "When they started [interviewing candidates] we knew they were doing it," Covey said, recalling that, at the same time, NASA was conducting glide-flight tests of the prototype orbiter, *Enterprise*.

So everybody's getting excited about the shuttle now. . . . We knew that NASA was getting ready. I had a vacation planned. I had just taken my wife and kids and put them on an airplane. They were on their way to California, and I was supposed to join them within a day or two. I got a call, and it was from Jay Honeycutt. Jay was calling to invite me to come to Houston. . . . It was very short notice for an interview. That was the first day they were calling anybody. Finally had got their list down and alphabetically they started calling people to come. I'm sitting there. I just sent my wife out. I'm supposed to go join her on this vacation out here. I remember thinking—I mean, this was the hardest question I was going to ask. I said, "Jay, so if I said I couldn't come next week, will you invite me back another time?" I later talked to Jay, and he said that he said, "Well, just a second. Let me check." So I go, "Oh, no."

Covey said that Honeycutt told him later that he had to go ask whether they could schedule another time for a candidate, since it was a possibility that hadn't been discussed. "They expect that everybody will say, 'I'll be there tomorrow,' you know. So he came back and says, 'Yeah, we'll invite you back.' Well, so I go on my vacation, and I'm going, 'Oh, my God. They haven't called me yet. When are they going to call me?' So it was a terrible vacation. It was a terrible vacation. Toward the end of it they finally called; said, 'Well, we're getting our stuff together. We want you to come week after next.'"

The interview process lasted a week and included physical, psychiatric, and psychological exams. "The physical exams included lab work of everything that they could measure," recalled Hauck. The psychiatric exam, Hauck said, involved interviews with a "good-guy psychiatrist" and a "bad-guy psychiatrist," each of whom played a different role in the test.

The bad-guy psychiatrist evaluated how you did under pressure. For example, "I'm going to read off a list of numbers. Tell me what they are in inverse order." And you start with two, five, and you say, "Five, two." And then three numbers, then four numbers, then five, then six, and you're sitting there just thinking, "I can't do this." At some point, you make a mistake. Inevitably, at some point you make a mistake and the psychiatrist said, "That's wrong," with a scowl on his face. "Can't you do better than that?" Blah, blah, blah. And, of course, he doesn't care whether you did it with five numbers, three numbers, or eight numbers. He's more interested in seeing whether you get flustered, wheth-

er you get antagonistic. And as I recall, I might have said, "That's the best I can do, yes." "That's okay."

The role of the other interviewer, Hauck said, focused more on the candidate's emotions and interpersonal relationship styles. "The good-guy psychiatrist would ask you questions such as if you were to wear a T-shirt and there were an animal on the front of the T-shirt and you wanted that to sort of be your symbol, what would that animal be? I forget what I said, and I'm sure he drew some conclusions whether you said a tiger or a turtle or a rat or what."

In another part of the test, he said, candidates were zipped up individually into a fabric sphere.

In order to get into it, you had to get into a fetal position, into a ball, and the concept of the sphere was it was just small enough so that it could go through the crew hatch in the Space Shuttle in the event that you had to rescue people from one shuttle to another. The charter was, "We're going to put you in this. You have oxygen. You have communications. We're not going to tell you how long you're going to be in there. At the end, we want you to write a flight report on what you think are the upsides, downsides, what more needs to be studied for this concept."

So that was fascinating. There was really two objectives there. One is, see how analytical you are about analyzing . . . a piece of hardware or software. Number two is a claustrophobia test, because you literally couldn't move very much, and it would be very clear if you had claustrophobic tendencies. As I recall, I found it most comfortable to sort of lie on my back with my knees up, and I almost fell asleep.

The "big deal" of the process, Hauck said, was the board interview with Johnson Space Center officials who made the selection decisions.

They'd say, "Tell us about yourself," and just let you talk. I don't remember getting any surprise questions, but some of the people got surprise questions. For example, President Carter was president at the time. He had just signed the bill that transferred the Panama Canal back to Panama, and one of the questions was, "What do you think about the Suez Canal situation?" And of course, the person might have started commenting about the Panama Canal because that was what was in the news, and then one of the board members might say, "Why are you telling us about the Panama Canal? We asked you about the Suez Canal." And again, it's an opportunity to see how people react under some level of stress and so on.

Interviewees were called to Houston in groups. John Fabian said his group comprised about twenty-two people, and he was convinced that any of them would have made fine astronauts. "It was all rather intimidating and awe-inspiring," Fabian said, "but somehow, at the end of it, some people got lucky, and other people didn't, and I was one of the lucky ones."

At six feet one, Fabian was too tall for earlier astronaut selections, but with the Space Shuttle program came a new maximum height of six feet four. Before that, he'd not given it much thought, he said. "I've always had the philosophy that you shouldn't try to be something you can't. I couldn't be an astronaut if I was six foot one, and that was above the height limit."

The highlight of the interview week for Terry Hart was the selection committee, led by Director of Flight Operations George Abbey, in part due to an unusual circumstance in another part of the interview. Hart's blood tests early in the week were flagged for being outside the parameters for uric acid.

"The basic message I was getting was that that was going to be disqualifying," Hart said.

And in a sense I think that really helped me, because I went into the interview just [like] I was down here for the experience and everything. I was relatively relaxed as you could be for such an interview and went through that interview process and finished the week up. I went home and told my wife that it was a wonderful experience, but I wasn't going to make it, which is what I thought from the beginning. But I was a little disappointed at that point, because as you get into the process, your competitive juices start flowing and everything. You really want to be part of this very exciting adventure that was about to begin. Yet realistically, I'd met all these people that . . . seemed to be so much more qualified than I was.

On the flip side, Norm Thagard found himself feeling like he was in the hot seat during his interview, particularly over a comment he made about women.

You sit there at a table and there are people on all sides of you during the interview and they're firing questions at you. . . . The question George [Abbey] asked me was, "Well, I see that you made a C in ballroom dancing. Why was that?" I said, "Well, our instructor was a woman who liked to lead." Which was true. "I found that very difficult to learn to dance with someone who was leading." But then the next question was, "Well, what do you have against women?" And, you know, they're firing these questions from all over and you're turning this way and then you're turning that way. [I heard] a little ruffle of movement and I see someone get up and leave. When I turn back, Carolyn Huntoon, who was the only female member on the thing, had gotten up and left. I said, "Well, this is just great." First of all, they've drawn this thing out, which to me, I thought was an innocent enough response, but now they're making a big deal out of it. Now this woman is obviously a feminist and offended that I've said this, and so she's left.

After the selection announcement, Thagard would find out what had happened when he and the other new astronauts were brought to Johnson Space Center for media events. "Carolyn Huntoon was the one that babysat our kids, because we brought them along for that," he said. "She took us in our car over to some of the events at JSC. I reminded Carolyn that she had gotten up and left during my interview and what I had thought was the reason why. She says, 'Oh, no, I had to get up to leave because my babysitter had to go home.' So it took me a long time to realize that, in fact, it hadn't been all that bad."

After the week-long interviews, it was time to wait. And wait. And wait. "I think I was [at Johnson for the interview] in August or something," Covey recalled. "And so it was go home and wait for five months to see what happened. And nothing; there was nothing. It was real quiet during that time period."

Finally, in January 1978 the phone calls started going out. John Fabian remembers where he was when he first heard that the new class had been selected, before he knew whether he had been chosen or not. "I was in bed that morning when my wife and I heard an announcement that NASA had selected thirty-five astronauts and among this group there would be six women," he said. "And my wife said, 'That's too many,' which sounds funny today. But, of course, her concern was that, if there are six women selected, that's six slots that my husband isn't going to fill."

It wasn't until later that day at work that Fabian got the phone call from NASA as he was preparing to go teach a class. "Mr. Abbey was on the other end, and he said, 'John, this is George Abbey; I'm calling from the Johnson Space Center. I'm interested to know if you're still interested in becoming an astronaut.' I said, 'Yes, I certainly am.' He said, 'Well then, I'm pleased to tell you that your name is on this list.' So I had to have somebody else go teach my class, because I was psychologically not prepared to go lecture at that particular time. It was a great thrill, a real honor."

At the time Mike Mullane got his phone call, he was stationed at Eglin Air Force Base in Florida but was on temporary duty to Mountain Home Air Force Base in Idaho. Mullane was phoned by his wife, who told him George Abbey had called their home to talk to him. Like several others, Mullane was aware the selection had been made because he had heard on the news about the women who were selected. Finally, he talked with Abbey himself and got the official word that he had been chosen. "I just went out and screamed with joy. I remember that night I bought some beer for the rest of the people that I was working with there at Mountain Home in the hangar there, and we had a little party. I remember . . . stopping out in the desert. This is out in Idaho. It's like New Mexico. Go out in the desert; it's like being in space. Black sky. I remember standing out there and just looking at the sky and thinking that I had this chance of actually flying in space."

Mullane said that, despite the news of his selection and that moment in the desert, he still had doubts that he would ever actually make it into orbit.

I'm one of these guys that tend to think of all the things that can go wrong, like a medical problem or the rocket blows up or whatever it is. . . . Even though Abbey called and told me that I'm an astronaut, I felt like there's still a lot that could go wrong that would prevent me from actually flying in space, but I still had this overwhelming sense of joy that I had this shot at getting into space. It was a lifetime dream come true to be an astronaut. But again, I didn't really ever consider myself an astronaut until the SRBs ignited on my first mission. All the rest of it I just thought it was name only. But it certainly was an overwhelming, joyful experience of the first magnitude.

We tend to set these goals and think that once we reach this goal, it's going to make you happy for the rest of your life. . . . Of course, that never happens. I remember telling my wife that if I just flew one time in space, just one time in space, that's all that I would need to be infinitely happy. And then I'll bet within two days after landing from my first mission, I said, "I sure would like another mission." It's just one of those things. It's a joyful experience to be told that you're going to get a shot at riding into space. So I was weightless at that point, I think. I was just floating around, already weightless.

Norm Thagard also had the yo-yo experience of assuming that finding out about the selection on the news meant that he hadn't been picked, only to get the phone call the next day at work from George Abbey saying that he had been chosen.

I hung up the phone and turned to the group that was there and said, "I guess I'm an astronaut." Then I went back in my room and put my head down on the desk and was real depressed for the rest of the day. That's honest. I was depressed. It took me awhile to figure out what was going on, but I finally think I understood that I'd always had goals, I always wanted to do this, that, and the other, but I never had really any goals beyond being an astronaut. So you're all of a sudden faced [with] there's nothing left to live for. Then you realize, well, yes, there is, because you still hadn't flown in space. So life goes on. But my reaction really surprised me at first, because it was depression. I mean, I was not elated at all. I remember feeling, on the one hand, sort of gratified, but on the other hand, just feeling real down.

The Thirty-Five New Guys were the first class of astronauts to come in as astronaut candidates, or AsCans. "In previous selections they had had some people that didn't really particularly care for the job and maybe didn't know what the job entailed and left," Mike Mullane said. "I think to avoid whatever embarrassment that might cause NASA or the individual, they established this plan which you come in for a couple of years and you go through training and evaluation. Then at the end of that period you either become no longer a candidate, and now you're an astronaut, or it's decided either mutually or by one party or the other that, yes, this probably wasn't the right move, so we agree to part company at this point. No hard feelings."

Mullane said he was afraid that the new process might mean that he would never earn the official title of astronaut.

I realistically thought there was a chance in a couple of years they might get rid of me. So I was concerned about that. I knew this was going to be an interesting mix of people, and I knew that there were going to be people that knew a lot more about stuff that was important than I knew, and there were going to be these pilots and all that other stuff, so I was a little concerned how we would all get along. But I think primarily I was just concerned about would I be able to really do the things that would be expected of me.

This first group of AsCans was so large that its members became the majority of the Astronaut Office when they reported for duty in July 1978. Loren Shriver recalled,

I think the folks who were still in the Astronaut Office, which, of course, had been between programs for several years of that period, were glad to see us there on the one hand, because they were all really busy doing the technical things that astronauts do while they're waiting to go fly—various inputs to boards and panels and safety inputs and crew displays and all that kind of thing. They were all really busy, and I think they were happy to see us show up so that we would be able to help them and take some of the load. At the same time, I think there was a bit of the "Oh, no, all these new guys. How are we ever going to get them trained and up to speed? Will they ever be ready to go fly in space?" Well, that's kind of a natural reaction to the group of people who has been there and done that a lot. That's a bit of a different aspect of "We're happy to have them here, but I don't know, it's maybe just a little more work for a while until we get them all checked out."

Dick Covey said that the remaining veterans were quite welcoming to the large surge of rookie astronauts. "I never felt like they saw us coming in as 'Oh, my God, we've got more people than we need,'" he said. "I've seen that since then, as the Astronaut Office has gone through huge swells and stuff, but I didn't sense that from them. I got the sense that the twenty something of them that were still in the office were looking forward to some additional help. We seemed to be welcomed very graciously, particularly by the [previous class]. They really embraced our arrival, and I always felt like they felt like they needed more people to do the work for the office and getting ready to fly."

The warm welcome was also experienced by Rick Hauck, who agreed that the "real astronauts" were grateful for the extra hands.

They'd already started gearing up for shuttle and they needed help. So we were there to be helpful in any way we can. They wanted to get us as smart about the systems as soon as they could. . . . Everyone was very hospitable to us, bending over backwards to make us comfortable and telling us how much they needed us. We felt wanted, and contrast that with Dick Truly and Bob Crippen, [who] had joined from the MOL *program, the Manned Orbiting Laboratory program, and when they arrived there, I forget whether it was Deke Slayton or someone else said, "We didn't ask for you. We didn't want you. Stay out of the way." Big difference. So I think that they were even sensitive to that kind of a reception.*

John Fabian said that he had also heard horror stories about how the last class—Group 7—had been treated when the recruits arrived in 1969. Unlike the other classes, Group 7 had not been chosen through an open selection. The air force had formed its own astronaut corps, independent of NASA, to support its Manned Orbiting Laboratory space station program. When that program was canceled, the air force closed its corps and asked NASA to take on its excess astronauts. At the time, NASA's astronaut corps had more people than it needed for the remaining Apollo-era seats available. There were reportedly multiple attempts at the JSC flight operations level to get rid of the new recruits, which were overruled by NASA leadership, eager to have the air force's support as the agency sought funding for the Space Shuttle.

"We heard some bad stories about the way the MOL guys were treated when they came in, as kind of a leper colony, and we weren't treated that way at all," Fabian said. "I think they were glad to see us come. The shuttle program was just around the corner, we thought. It turns out it wasn't quite just around the corner, but we thought it was, and there was a lot of work to be done, and there was a lot of legwork that needed to be accomplished. . . . So I think they were glad to see us. They got some new hands and legs, and I think that they counted on us being somewhat motivated and somewhat capable. So it was a very pleasant thing."

Steven Hawley was a little less sure what the veterans thought of his AsCan class.

We hadn't really been flying a lot in '78; . . . since we'd landed on the moon, there'd only been like four crews to get to fly, and here's this new bunch of guys walking in the door. I could see how some of the guys that had been around for a while waiting to fly might have been a little resentful. If they were, that didn't come across in any way, because our training was separate from what most everybody else was doing. Everybody else was doing mainstream support of shuttle and development of everything that needed to be done before STS-1. We would cross paths at the Monday morning meeting or you'd run into them at the gym or something like that, but mostly we did our own thing.

Initially, Hawley said, the main interaction between the new class and the veteran astronauts came in the form of NASA history lessons during their training. "They thought it was important, and I think it is, that we hear from people that had flown Apollo and had flown Skylab and ASTP. So I remember we got lectures from some of the guys that were still there, some of the guys that had left but came back to talk to us about their flights and what it was like back then. . . . These were my heroes, and to actually get to sit in a room and listen to them talk about their flights was pretty awesome."

Veteran Apollo astronaut T. K. Mattingly confirmed that the corps was glad for the new class and the needed help it provided.

Once we got these folks on, the OV-101 [Enterprise] *was rapidly approaching the time to get ready to go. So we put together the training program for the new folks and helped them get started on that. Then we split them up, . . . just spread amongst the few of us that had been around. [The shuttle's robot arm] and a lot of these other activities were all getting sort of a lick and a promise instead of real attention till the '78 group came on board, and once they went to work, then they really took hold and played very key roles in the development.*

Of course, the "Thirty-Five New Guys" weren't just "guys." The veteran flyboys of the astronaut corps also welcomed for the first time six women who were part of the new class. Sally Ride pointed out that her class presented a double whammy for those who had already been in the corps—not only were they now outnumbered by rookies, they were the minority in an office that was suddenly much more diverse. "They seemed to accept us pretty well," recalled Ride.

We had them outnumbered, so I'm not sure they had a choice. It was clearly very different for them. They were used to a particular environment and culture. Most of them were test pilots. There were a few scientists, but most were test pilots. Of course the entire astronaut corps had been male, so they were not used to working with women. And there had been no additions to the astronaut corps in nearly ten years, so even having a large infusion of new blood changed their working environment.

But they knew that this was coming and they'd known it was coming for a couple of years. Well before the announced upcoming opportunity to apply for the astronaut corps, NASA had decided that women were going to be a part of it. So I think that the existing astronauts had a couple of years to adjust and come to terms with it. By the time that we actually arrived, they had adapted to the idea. We really didn't have any issues with them at all. It was easy to tell, though, that

9. The first female astronaut candidates in the U.S. space program, *left to right*, are Sally Ride, Judy Resnik, Anna Fisher, Kathryn Sullivan, and Rhea Seddon. Courtesy NASA.

the males in our group were really pretty comfortable with us, while the astronauts who'd been around for a while were not all as comfortable and didn't quite know how to react. But they were just fine and didn't give us a hard time at all.

Ride said there was a lot of media attention surrounding the TFNG announcement since it was the first astronaut selection in ten years and the first selection to include women. She said the attention didn't affect her private life all that much, since the agency worked to keep the extra attention at a minimum so that it didn't affect the astronauts' abilities to train and work. "It wasn't particularly burdensome after the initial flurry of interviews," Ride said. "There was a fair amount of [attention], but it was still easy to have a normal life. . . . I think JSC worked hard to prepare for the arrival of women astronauts and female technical professionals. The technical staff at JSC—around four thousand engineers and scientists—was almost entirely male. There was just a very small handful of female scientists and engineers—I think only five or six out of the four thousand. The arrival of the female astronauts suddenly doubled the number of technical women at JSC."

Joe Engle, who became an astronaut in 1966 as part of the "Needless Nineteen," recalled that one dilemma with adding women to the corps was figuring out where to put the women's locker room at the gym. "We just had a guys' locker room over there up to then," Engle said. "So Deke [Slayton] had to figure it out. And of course, good old Deke, he said, 'Well, hell, we'll just put a curtain up and you can all use the same damn room,' but he finally conceded that they would have a separate room."

For some, this was their first time to ever work closely with women. "It was really all new to me, and I didn't do it all right, either," Fabian recalled. "That was a slightly different era. It was an era in which you would take the centerfold of *Playboy* magazine and post it up on the back of your office door, and that was thought to be totally acceptable as long as it was the back of the door instead of the front of the door. And people hadn't yet thought of the word 'harassment.' We were all learning. We were all learning in those days."

Mike Mullane said that working with his female classmates was a new experience for him as well. "I'd be a liar if I didn't say it was difficult to learn how to work with women," he said,

and not because of the women; because I had no life experience in working with women. I tell everybody, there were two things that at age thirty-two I did that I had never done in my life, when I woke up to go to work for my first morning as an official astronaut at NASA*. . . dressed myself, and worked with women.*

I went to twelve years of Catholic schools, wore a uniform every day. Woke up, put on a uniform. Went to school. Went to West Point. For four years I don't think I ever saw an article of civilian clothes. Didn't have it in the closet. Wore a uniform all the time. Went into the air force. Would wake up in the morning, go to work, put on a flight suit. Not one time in my life did I ever have to go to a closet, open it up, and pick a pair of slacks and shirt that matched. And that was a real struggle. In fact, a number of times that I walked out of the house or walked through the kitchen on my way to work, Donna, my wife, would look at me and say, "You're not going to work dressed like that, are you?" In fact, she told me she was going to get Garanimals and put them on the clothes so that I could match the elephants with the elephants and the giraffe with the giraffe.

Mullane said that unfamiliarity with dealing with civilian clothes was a common struggle among the career-military astronauts. "I wasn't the only one struggling in this regard, because I remember driving up one day to NASA with my kids in the car, . . . and there was one of the astronauts walking around in plaid pants," he said. "Plaid pants. I mean, even I, with my absolute zero fashion sense, thought that maybe that looked a little bit retro. In fact, to this day, my kids, they're in their thirties now, if I'm with them and they see a golfer out in plaid pants, my kids will laugh and say, 'Hey, Dad, check it out. There's an astronaut.'"

Just as he'd never been in an environment where he didn't wear a uniform, Mullane said, those environments also limited his contact with the female sex. "I had never in my life ever worked professionally with women," he said.

In fact, my whole life, the Catholic schools I went to weren't gender-segregated, but a lot of the classes were. . . . So I had very little interaction with females as a young person, and West Point had no females at the time I was there. In the air force, the flying community that I was in had no females when I was in there. So as a result, I was thirty-two years old when I was selected as an astronaut and I had never worked professionally with women, and I have to admit that I'm sure I was a jerk, in a word, because I just didn't know how to act around them, telling jokes that probably were not appropriate to tell and just doing dumb things that were inappropriate and probably would have gotten me a prison sentence in this day and age now with sexual harassment and all that. The women had to endure a lot, because there was a lot of guys like myself in that regard, I think, that had never worked with women and were kind of struggling to come to grips on working professionally with women, but we all made it. That's for sure.

The camaraderie among the thirty-five was interesting, too, because of the extreme diversity and the introduction of the new mission specialists. There was a difference even in the aspirations that led each new astronaut to the corps. The TFNG class was announced almost nineteen years after NASA hired its first astronauts, and several of the pilot candidates in Group 8 had spent much of their career with the idea in the back of their minds of someday joining their ranks.

One of those was Dan Brandenstein, who had been interested in becoming an astronaut for a long time and had set himself milestones to accomplish to get there. "But a number of the mission specialists, they weren't pilots and they never had been pilots. I think Sally Ride is one that comes to mind. I mean, she was saying how she was just walking through the Student Union one day and saw a flyer that said NASA was looking for astronauts, and that's really the first time she'd ever thought about it. A number of the mission specialists, that was their attitude."

To Brandenstein, the new diversity his class brought to the astronaut corps was a good thing. Like others, he'd been in much more homogenous environments until then and was excited about having his horizons broadened. "The wide diversity of backgrounds that we had in that class was unique to NASA, and I personally loved it, because I've always been interested in a lot of things," he said.

I'm fascinated going into a factory where they make bubble gum or you name it, just to see how different machines and different things work. In my lifetime, I took up skiing and I didn't take lessons; I learned to do it through the school of hard knocks. I bought a sailboat and I made some sails because I thought it would be kind of fun to make a sail. So I was always interested in not just what I did, but kind of a wide variety of things. So being now in a group with people that were doctors and scientists and all, this was really fascinating to me.

Mattingly said the new mission specialist category of astronaut required major adjustments that the Astronaut Office pretty much made up as they went along. For example, he said, he was involved in some discussion about who should command the Space Shuttle missions. Up to that point, only four NASA missions had included nonpilot astronauts—four scientist-astronauts had flown over the course of the last moon landing and the three Skylab missions. On those flights, the commander was still chosen from the pilot astronauts in the corps. Mattingly, who said that his experience on aircraft with larger, more diverse crews gave him a different perspective, wondered whether that tradition should change with the new mission specialists.

We came up with these crazy ideas that since we're going to be flying this "airplane," but the mission of the airplane is whatever is in the payload bay, maybe the mission commander should be a mission specialist, or maybe the mission commander is a separate position where both pilots and mission specialists aspire to that being the senior position. The skipper of a ship doesn't put his hands on a steering wheel; he directs the mission. I thought that was really good, and some of my navy buddies, "Yeah, that makes sense." Boy, that did not float at all, and there was a bigger division between mission specialists and pilots than I had ever guessed there would be.

Mattingly intermixed pilots and mission specialists on the teams developing the Shuttle Avionics Integration Laboratory (SAIL), thinking the designations wouldn't matter much. "I just mixed them up," Mattingly recalled.

I said, you know, "Bright people work hard. I don't care where you go." So we sent mission specialists and pilots both to the SAIL, and the job that you had to do over there didn't require any aeronautic skills at all; it was checking out the software and just going through procedures. Anybody who was willing to take the time to learn the procedures and has some understanding of how this computer system works is going to be fine.

We ended up having to put out all kinds of brush fires, you know, "He can't do that. He's a mission specialist." . . . We had a SAIL group around the table when they were having a debriefing. We did this every week to go over all the things we'd done and what was open. Steve [Hawley] started it off with, "Did you hear about the pilot that was so dumb the others noticed?" I've told that to a lot of people, and I thought that was great. And at that point I think that Steve finally broke the ice, and everyone kind of said, "This is dumb, isn't it." After that, at least it never came to my attention again if they had any problems, but from then on, they really came together.

Brewster Shaw recalled that early on Kathy Sullivan commented to him that the pilots were going to be just like taxi drivers and that it was the mission specialists who were going to do all the significant work on the Space Shuttle program. "Turns out, by golly, she was pretty much right," Shaw said. "But at the time, being a macho test pilot, I was a little appalled at her statement." Shaw saw just how true Sullivan's statement was as pilot of STS-9, his first mission. "Our role was very minimal, John [Young's] and mine, because we didn't have to maneuver the vehicle very much, but we had to monitor the systems a lot. So we didn't get to participate to a great length in the science that was going on. So, yes, pretty much, we got it up there and we got it back. In the meantime, the other guys did all the work."

For training the TFNG group was split into two sections—the Red Team and the Blue Team, Brandenstein explained. Teams took turns in the classroom and flying, spending a half day on each activity and then switching off with the other team. "We broke in two parts because the classroom didn't hold the thirty-five people," John Fabian recalled, "and so you got to know the people that were in your class much better, in reality, than you got to know the people that were in the other class, particularly in the first three or four months. But what we found out very quickly was that all of these people, whether they were the youngest in the group or the oldest in the group, they were all extraordinarily bright, extraordinarily capable, and very, very eager to succeed in what it was that they were doing."

The new AsCans brought a variety of interesting backgrounds to the table, according to Brandenstein.

There were a lot of neat people . . . that knew a lot about things I didn't have a clue about, so you could learn a lot more. And that was kind of the flavor of the training. The first year of training, they try and give everybody some baseline of knowledge that they needed to operate in that office, so we had aerodynamics courses, which, for somebody who had been through a test-pilot school, was kind of a "ho-hum, been there, done that," but for a medical doctor, I mean, that was something totally new and different. But then the astronomy courses and the geology courses and the medical-type courses we got, all that was focused on stuff we'd have to know to operate in the office and at least understand and be reasonably cognizant of some of the importance of the various experiments that we would be doing on the various missions and stuff. So I found that real fascinating.

For example, he said, the astronomy course was a real standout for him in the training. "He's passed away now, but the astronomy course was Professor Smith out of University of Texas, and he was kind of your almost stereotypical crazy professor," Brandenstein said.

I mean, he was just a cloud of chalk dust back and forth across the blackboard as he went on. We had twelve hours of astronomy; he claimed that he gave us four years of undergraduate and two years of graduate astronomy in twelve hours. And it gave you a good appreciation of what it was all about. It didn't, by any stretch of the imagination, make me an astronomer, but the intent of it was to give you an appreciation and give you an understanding. And then also because of the very special instructors they brought in, it gave you a point of contact, so if somewhere later in your career you had a mission that needed that expertise, you had somebody to go up to and get the level of detailed information you needed.

Brandenstein said that he was very grateful that the format of the classes was just to absorb as much as possible from the barrage of information; there were no written tests. "Everything was pretty much that way. It was just dump data on you faster than you could imagine. A common joke was that training as an astronaut candidate was kind of like drinking water out of a fire hose; it just kept coming and kept coming and kept coming. Probably the good point of it was you weren't given written tests, so they could just heap as much on you, and you captured what you could. What rolled off your back, you knew where to go recover it."

Everybody in the class had their strengths at some point in the training, and Brandenstein said the pilots enjoyed when the training was on their home turf. The pilots in the group, he said, liked to take the mission specialists up for what they referred to as "turn and burn"—loops, rolls, and other aerobatics. "We'd go out and do simulated combat and show them what it's like to have a dogfight and all those sorts of things. So that was [as] fascinating to them as me sitting down with an astronomer or a doctor and finding out about the types of things they did."

Brewster Shaw said he came into the corps with very few expectations of what the job entailed, beyond eventually flying on the Space Shuttle. "I soon learned that the percentage of the time you got to fly the Space Shuttle was pretty miniscule, relative to the percentage of the time that you were here working for the agency, and that there was a lot of other things you were going to do that would take up all your time, and that was made clear to us pretty soon."

4. Getting Ready to Fly

One downside of creating something from scratch is that it doesn't come with an instruction manual. As shuttle hardware development was winding down, astronauts began facing the next challenge—how do you teach people to fly a vehicle that no one has ever flown? Step one was figuring out how exactly the thing would fly and then developing procedures for flying it. That would be followed by step two, developing procedures for training people to follow the procedures for flying.

As the people who would have the job of actually flying the as-yet-unflown vehicle, the astronaut corps was tasked with providing operator input on the creation of both operating and training procedures. Since a large percentage of the corps at the time was rookie astronauts, in many cases procedures for flying a vehicle that had never been in space were being developed by astronauts who had also never been in space.

For example, the team of astronauts representing the Astronaut Office in figuring out how to use the shuttle's robot arm, the remote manipulator system, was composed of astronauts John Fabian, Bill Lenoir, Bo Bobko, Sally Ride, and Norm Thagard. "One of the things that we worked on early," recalled Fabian, "was the failure modes of the robotic arm and how to protect the orbiter and the crew in the event one of these failure modes occurred. At the same time, we tried to figure out ways that we could continue on and do the job and do the mission at hand in the face of certain failures of the robotic arm."

To tackle these issues, the team ran many simulations in Canada, where the arm was made, to work through procedures for nominal and off-nominal operations. Fabian recalled staying in Toronto quite a bit, around four hundred nights.

I used to try to change hotels about every one hundred nights up there because it got kind of old, staying in the same motel. And by far, I spent the most time up there of the people that were working on it, because I was there in some very, very long simulations, trying to figure out the various workarounds. Some days were very long. Some days were seven o'clock in the morning until nine or ten at night, and we would take a break and go out and have dinner and then maybe go back to the plant. This wasn't common. Usually it was close to an eight-to-five job, or a nine-to-six job or whatever. But if you were in the middle of a long simulation, then you would work longer hours.

Fabian recalled that it was Sally Ride who wrote the procedures for the first flight of the arm on STS-2. The procedures, Ride said, came directly out of the simulations.

Until you actually start using something, it's very difficult to make predictions on how well it's going to work, what it's used for, and how to accomplish the tasks that it's designed to accomplish.

We did a lot of development of the visual cues. The astronaut controlling the arm looks at it out the window and also monitors its motion using several cameras. Often critical parts of the view are blocked, or the arm is a long ways from the window, or the work is delicate. In those cases, the astronaut needs reference points to help guide the direction he or she moves the arm. How do you know exactly that you're lifting a satellite cleanly out of the payload bay and not bumping it into the structure? We also helped determine how you move the arm. What limits should be put on the use of the arm to make sure that it's kept well within its design constraints? We did a lot of work on that. It was rewarding work, because it was at a time when the system was just being developed, and nobody had paid attention to those things yet.

Despite challenges like limited visibility, Fabian said that the work developing the interface and procedures resulted in the manipulator being "quite easy to use." But, he noted, "it's also a little bit intimidating, because you've got this thing which is fifty feet long out there in the cargo bay, and if you're not careful, you could punch a hole in the wing or do something really stupid with it."

Fabian views his contributions to RMS development as one of his most significant accomplishments at NASA.

The RMS has worked wonders on all of the flights. It's really a great piece of equipment. The Canadians have gone on to design an arm now for the Inter-

national Space Station, so they have gotten a big return out of their early investment in developing these electromechanical systems. We really had an opportunity with the RMS to work on the human interface, to make it something which is straightforward and easy to use and intuitive in its application. That's now followed over into the Space Station, and potentially it will go on to other applications. I think it's the most significant thing that I did in my time, and I think it's the thing I'm proudest of.

Fabian, along with astronaut Judy Resnik, was also involved in establishing how crew duty assignments would work during the Space Shuttle program. Previous NASA missions had carried crews no larger than three astronauts, and each astronaut had his own title and assignment. With the larger shuttle crews, there would be multiple mission specialists, and NASA had to decide how crew duties would be allocated and assigned. The system that was developed assigned mission specialist (MS) 1 overall responsibility for payloads and experiments in orbit. Mission specialist 2 was given primary responsibility for flight engineering, helping the pilot and commander during ascent and entry, and serving as backup for payloads. Mission specialist 3 had responsibilities for independent experiments and extravehicular activity, or EVA.

"MS3 would be typically the most junior and the lowest training requirement but heavy on EVA," Fabian recalled. "MS1 would have the largest overall responsibility and, in principle, ought to be the most experienced member of the astronaut mission specialist crew. And MS2 had the most simulation time and spent an enormous amount of time in the simulator. . . . We split it that way in order to recognize the fact that the flight engineering role was the dominant training requirement for one of the mission specialists, and therefore that person shouldn't be burdened with overall responsibility for the satellites."

Astronaut input was also crucial in the development of the Shuttle Motion Simulator, or SMS, which would become the primary system for training Space Shuttle crews. The SMS was the only high-fidelity simulator capable of training crews for all phases of a mission, beginning at T minus thirty minutes and including simulated launch, ascent, abort, orbit, rendezvous, docking, payload handling, undocking, deorbit, entry, approach, landing, and rollout. The simulator could tilt up to ninety degrees, to the vertical position the orbiter would be in during launch, and could simulate the vibrations and noises of ascent.

Bryan O'Connor, who was selected as an astronaut in 1980 and assigned to SMS development, described what it was like to see the more veteran Apollo-era astronauts apply their experience to the project.

All these guys had quite a history back through the Apollo program, and it was difficult not to pick up some of that climate and the cultural aspects of that, the pride that they had in that program, the frustrations they were having as we went forward, and things not being the same as they had been before, where it seemed like there was plenty of money. Now the environment we were in was a little different, but a lot of the cultural aspects that had made the Apollo program great were interesting to me to jump into and start learning about.

Since the Space Shuttle had yet to fly at the time, the development of the SMS was based on analysis and ground-test data. "Part of what they were doing was to try to make things like the visual cues and the oral cues that they have for the crew as accurate as they could so that the environment in that trainer would be as close to real as they could get it," O'Connor recalled.

I would be in the third seat taking notes and we would have two Apollo guys like John Young and T. K. Mattingly . . . trying to remember what it sounded like when the reaction control jets fired. The engineers would be outside the simulation putting in these models and turning up the volume and changing the pitch and the frequencies of these noises in the cockpit to make them sound space-like. Nobody knew what it would sound like on shuttle, but if they could make it sound something like what Apollo sounded like, they thought that was a good start. So it was fun to hear these guys arguing about whether some noise that was in there was accurate to the Apollo sounds when neither one of the two guys in the front had actually flown on the Apollo system for some years.

Determining how to calibrate the vibrations in the SMS was another instance where O'Connor recalled the Apollo veterans really having a good time.

There was a fellow named Roger Burke, who was the head engineer in charge of developing the Shuttle Mission Simulator, and Roger had a little bit of a diabolical streak in him. One day we were trying to simulate what the vibrations might feel like during launch. Roger was asking these pros—and I think John Young was in the commander's seat on this one—for advice on how much vibration seemed right. So we did several launches in a row and each time John would say, "No, you need some more. It's going to vibrate more than that." So you can picture Roger Burke out there, turning this potentiometer up to get more vibration in there. Then we would fly another launch and John would say, "Nope, nope. That old Saturn, that had a lot more vibration than that. You're going to have to tweak it up a little more, Roger."

So after about three or four of these things, Roger decided he was going turn it all the way to the max, and he did, and that was one hellacious ride. John and I both knew what he had done, and all we could do was just hold on. We were strapped in, but you still felt like you needed to hold on. You couldn't see any of the displays at all. It was just a big blur, and we were bouncing around like it must have been the case in the old days, when people were going down the rutted dirt highways in buckboard wagons or something. Roger actually broke the system on that particular run.

O'Connor said that his role in the development of the SMS taught him a valuable lesson that stayed with him throughout his time at NASA. "As I look back on it, I realize that what I thought was going to be a terrible job and not very much fun and out of the mainstream was just as important as any other job anybody else was doing. It was kind of an early learning event for me, because I realized then, and it came back to me many times later on, . . . that everybody's important, no matter what their job is in the space program. There aren't any nonimportant jobs."

In regard to simulators like the SMS, STS-1 pilot Bob Crippen noted that after the shuttle started flying, astronauts brought their experience back to the simulators to make them more accurate. "I know on the first flight we learned that the reaction control jets, which are used to maneuver the vehicle while you're on orbit, really are loud," Crippen said. "It sounds like a Howitzer going off outside the window. . . . I knew they'd be loud, but it was louder than I expected. It was a good thing to at least try to simulate that a little bit better in the simulator so that people weren't really surprised by it."

While ground-based simulators were used to train astronauts on the launch experience, a modified Gulfstream jet was used to re-create the experience of landing. The exterior of the Shuttle Training Aircraft (STA) was modified to better withstand the stresses of replicating the shuttle's reentry profile. Changes to the cockpit echoed the layout of the shuttle's flight deck and the view that astronauts would have while landing the shuttle.

Engine thrust reversers reproduced the "flying brick" aerodynamics of a gliding shuttle.

"The Gulfstream was extremely valuable," said astronaut Joe Engle. "It was a very, very good simulation, a very accurate simulation of what the orbiter would do. The response was tuned as data would come back from the orbiter flights, but it was a very, very good training airplane and still considered by pilots as the very best single training tool that they have to land the shuttle."

Astronaut Charlie Bolden praised the STA for giving him an excellent feeling for how the shuttle handled during entry and landing. "When I was Hoot's [Gibson's] pilot, sitting in the right-hand side, calling off airspeeds backwards, airspeed and altitude, putting the landing gear down, it was just like being in the STA. Combine that with the SMS, it was just as if I had been there before. So, the world of simulation, even back then when it wasn't as good as it is today, was awesome."

Of course, an even better tool than an airplane for learning how an orbiter lands is an orbiter. In an unusual turn of events, the Space Shuttle landed before it ever launched for the first time.

The first orbiter completed was *Enterprise*, a flight-test vehicle ultimately incapable of flying into space but fully functional for atmospheric glides. Originally, *Enterprise* was to be a member of the operational orbiter fleet, a fact reflected in its official vehicle designation. Each of the operational orbiters has an Orbiter Vehicle, or OV, number. *Enterprise*, the first of the fleet, is OV-101. *Enterprise* was to be joined by three other orbiters: *Columbia*, OV-102; *Discovery*, OV-103; and *Atlantis*, OV-104. (Later would come *Endeavour*, OV-105.) An additional orbiter, the Static Test Article, was built as a ground-based test vehicle and was given the designation STA-99. *Enterprise* was the first of the flying orbiters to be completed, but it was not given a space-worthy configuration. Rather, it was modified for use in glide flights to test how the vehicle would fly during entry and landing. The vehicle was also used for vibration testing at Marshall Space Flight Center in Huntsville, Alabama, to determine how an orbiter would withstand the stresses of launch, and to test the facilities at Kennedy Space Center in Florida to make sure they were ready for processing vehicles for launch. The plan was that all of those tests would be completed during the mid-1970s, and then *Enterprise* would be modified to take its place alongside *Columbia* as a spacecraft. However, it was determined that because of the weight of *Enterprise*, it would actually be easier to modify the Static Test Article instead, and thus STA-99 became OV-99, *Challenger*, while *Enterprise* would be used for other ground tests and ultimately would be the first orbiter to become a museum exhibit. Since the initial construction of OV-101 was completed in 1976, it was originally to be named *Constitution*, in honor of the U.S. bicentennial year. The name was changed as the result of a letter-writing campaign to the White House organized by fans of the television series *Star Trek*, which featured a starship named the USS *Enterprise*.

Between February and November 1977, *Enterprise* made a series of test flights, dubbed approach and landing tests (ALTs). The first flights were captive-carry tests, in which *Enterprise* was carried, unmanned, on the back of a specially modified Boeing 747 to study its aerodynamics. For later flights, the orbiter was released from the 747 during flight and piloted to the ground. Assigned as *Enterprise*'s crews for those flights were astronauts Dick Truly, Fred Haise, Gordon Fullerton, and Joe Engle.

Just prior to the start of the ALT program, Fullerton was a key player in the design of the orbiter cockpit, which he believes contributed to his selection for the ALT program. "I'd run across a lot of really crummy designs in learning to fly certain airplanes, and I thought I could do better," Fullerton recalled.

As it turned out, that was a real challenge. With the shuttle, rather than lying on your back on the end of a rocket riding into space, you had possibility of controlling it, both in the vertical mode and coming back as an airplane pilot at the end. The whole complexity of it is far more complex than the [earlier NASA] rockets, as far as what the man could do. Putting all that together in a cockpit was really intriguing, and I enjoy working with stuff in an engineering sense, so it was perfect. I became the cockpit design czar, sort of, to go to really organize and set up and go to all the reviews. I had a big foam-core cardboard mockup of the entire cockpit built right there in the Astronaut Office, and I cycled all the other guys in there to say, "What can you see? What would you do if this was a checklist? Can you reach it?" So I did a human factor study on all that.

With *Enterprise*, Fullerton saw the designs and drawings he had signed off on come to life. "It's very satisfying when you see [the results of what you did]," he said. "I can go get in an orbiter right now and look at the panels and think, 'Oh, yeah, I remember all this.' It's a real feeling of personal pride."

10. An aerial view of *Enterprise* hoisted into the Dynamic Test Stand at NASA's Marshall Space Flight Center for the Mated Vertical Ground Vibration test. The test was the first time that all of the Space Shuttle elements were mated together. Courtesy NASA.

As with the shuttle itself, training for the *Enterprise* test flights was very much a make-it-up-as-you-go-along process, Fullerton explained. "People say, 'How do you train?' thinking, well, you go to a school and somebody tells you how to do it. It's not that at all. Somebody's got to write the checklist, so you end up writing the checklist, working with each subsystems person and trying to come up with a prelaunch checklist for the approach and landing tests. So you're doing the work, and the learning comes from doing jobs that needed to be done."

The ALT flights blurred the distinction between the crew members' new careers as astronauts and their past experiences as test pilots. "[For] astronauts now, the orbiter's a pretty stable configuration, so they go to a school with ground school instructors that know the system. . . . For ALT and then subsequently on the *Columbia*, we were clearly test pilots because we were doing stuff that there wasn't a procedure for. We were writing the procedure and then flying it for the first time."

Not only did procedures have to be developed for the test, NASA had to decide exactly what the test would look like. On the one hand, for safety reasons, testing would have to be incremental, starting with the relatively low-risk captive-carry flights and ending with actual flight-profile landing tests. From that perspective, more flights were better, to assure safety at each step and to maximize the data results of the tests. On the other hand, the ALT flights would use money and manpower that would otherwise be going toward preparations for the actual spaceflights of the shuttle, so more tests would delay the first flight. After a lengthy debate, the decision was made to not pick a number in advance, but to decide in real time what was needed. Ultimately, thirteen ALT flights would be made.

"There were five captive, inert flights," Fullerton explained,

where the orbiter was bolted on, completely inert, nothing moving, nothing running other than some instrumentation. Fitz Fulton and Tom McMurtry and flight engineers flew those five to the point where they said, "Okay, the combination is clear, and we understand what we've got here." So then they decided to have some x number of captive, active flights, where the crew got on board and powered up the APUs and the electronics and all the subsystems, and those were dress rehearsals up to launch point. They had an open number of those. It turns out after three, they thought they'd learned all they needed to know. The systems were working. Had a couple of failures on number two, a big APU propellant leak. I was chasing that one. At three, they said, "Okay, it's time to go do it," and they were trying to get to the end as quick as possible, so they could get on with the Columbia.

After the three crewed flights atop the 747, it was time to set *Enterprise* free and see how the vehicle and its crew would do on their own. Fullerton was pilot and responsible for monitoring the orbiter's systems. With him, as commander, was astronaut Fred Haise. "On the very first flight, the instant we pushed the button to blow the bolts and hop off the 747, the shock of that actually dislodged a little solder ball and a transistor on one of the computers, and we had the caution tone go off and the red light. I mean instantly," Fullerton recalled.

I'm looking, and we had three CRTs, and one of those essentially went to halt. It was the one hooked to one of the four computers that monitored. This is pretty fundamental. All your control of the airplane is through fly-by-wire and these computers. So I had a cue card with a procedure if that happened, that we'd practiced in the simulator, and I had to turn around and pull some circuit breakers and throw a couple of switches to reduce your susceptibility to the next failure. I did that, and by the time I looked around, I realized, "Hey, this is flying pretty good," because I was really distracted from the fundamental evaluation of the airplane at first.

Haise recalled that it was surprising to him not to be able to see the 747 beneath him while *Enterprise* was still attached to the airplane. "No matter how you'd lean over and try to look out the side window you couldn't view any part of it, not even a wingtip. It was kind of like a magic carpet ride. You're just moving along the ground and then you take off. It was also deceptive sitting up that high. Things always looked like it was going slower than it was, for your taxiing and particularly the first takeoff, it didn't look like we were going fast enough. I said to myself, 'We're not going fast enough to make it off the ground.'"

Leading up to this point, predictions and models had proposed how an orbiter would fly. There had been simulations, both in training equipment on the ground and on the Gulfstream in the air. But now it was time to move beyond the models and simulations. For the first time, NASA was about to learn how the orbiter would fly from the real thing.

"To me, even at the first flight, it was very clear it handled better in a piloting sense than we had seen in any simulation, either our mission simulators or the Shuttle Training Aircraft," Haise explained.

It was tighter, crisper, in terms of control inputs and selecting a new attitude in any axis and being able to hold that attitude, it was just a better-handling vehicle than we had seen in the simulations, although they were close.

The landing also was a pleasant surprise from the standpoint of ground effect. Ground effect is a phenomenon you run into. When you get within one wingspan height of the ground, you start running into air-cushioning effects,

11. The Space Shuttle *Enterprise* participating in approach and landing tests. Courtesy NASA.

which can, depending on the vehicle's shape or configuration, be very different. It turned out the shuttle, in my view, was a perfect vehicle. If you get set up with the right sync rate, coasting along, you can literally almost go hands-off, and it'll settle on and land itself very nicely.

Flying two of the *Enterprise* flights—the second and fourth—were Joe Engle as commander and, as pilot, Dick Truly. Engle and Truly went on to crew the second shuttle flight, STS-2. Engle had participated in similar flights as part of another space plane program, the X-15, a joint endeavor by NASA and the air force, and had actually earned astronaut wings for flying into space with the air force before being brought in to NASA's astronaut corps. As with the *Enterprise* test flights, the X-15 began its flight by being lofted by a larger aircraft. "I think one of the reasons that I was selected to fly the shuttle, initially," Engle said, "was because of the experience that I'd had at Edwards (Air Force Base) with the X-15 and air launching from another vehicle, from a carrier vehicle."

The orbiter's aft section was covered with a tail cone on all of the captive flights and on the first three free flights. The tail cone reduced aerodynamic drag and turbulence and was used on all flights where the orbiter was ferried cross-country. The last two free flights were flown without the tail cone, thus exposing the three simulated Space Shuttle main engines and two Orbital Maneuvering System (OMS) engines, and most closely simulating actual conditions. The tail cone would continue to be used any time an orbiter was transported atop a 747 for upgrades or after a landing, up to and including their final flights to museum homes in 2012.

The tail cone not only made the orbiter easier to fly, it increased the time of the glide flight by reducing drag and slowing the vehicle's descent rate. While the longer glide time—about five minutes with the cone versus about half that without—allowed NASA to gain more data on the orbiter systems in flight, the most important data would come when the cone was removed.

"The orbiter flew pretty benignly with that tail cone on," Engle explained. "But that was not the configuration that we needed to really have confidence in, in order to commit for an orbital launch. So although we were getting more time on those systems with the tail cone on, from a performance standpoint and a piloting task standpoint, we really didn't have what we needed until we flew it tail cone off."

The fourth ALT flight—the first without the tail cone—flew October 12, 1977. By that time the decision had been made that the program would end with the fifth free flight. While the first four landings had been on a dry lake bed, the final flight would land on the main concrete runway at Edwards Air Force Base. To ensure successful accomplishment of that more-challenging landing, the flight would be focused solely on that task. As a result, any last in-flight research would have to be conducted on the fourth flight.

For Engle, the development of the profile for that fourth flight was the most exciting and more rewarding time in the ALT program. "Our goal on our flight was to pack as much meaningful flight-test data as we possibly could in that short period of time," he explained. "We did work very hard, not only on the simulators, but at Edwards in both T-38s and the Gulfstream aircraft, in going through and tailoring and modifying and readjusting the profile so that we literally wasted no time at all from one data point to the other. We would go from one maneuver and make sure that we were set up for the subsequent maneuver. It was a very demanding, fun task."

The grand finale of the ALT program was free-flight five, with Haise and Fullerton making the first landing on a concrete runway. While largely successful, that flight test led to the discovery of a flaw in the design of the flight control software that led the pilot into a pilot-induced oscillation. "We bounced around and shocked a lot of people, probably more than [we realized]," Fullerton said. "It didn't look that bad from inside the cockpit. But, again, that's why you do tests. You find out. Then the debate was, should we fix that and test it some more—it was a strong feeling that was a pretty exciting landing, which shouldn't be that exciting—or do we cut it off, fix it by testing and simulators, both airborne and on the ground. Do we know enough to press on? And it turned out that was the decision: you've got to cut the ALT off so we can go on the *Columbia* and get into orbit."

While the off-nominal landing—and the control problems that caused it—produced some concern, the experience was not without a silver lining. While most people had been pleased that, as a whole, *Enterprise* had flown and landed better than the models had predicted, the softer-than-predicted landings caused a problem for one particular team.

"The landing gear people were somewhat chagrined through most of that test program, because we were not landing hard enough to get them good data for the instrumentation they had on the landing gear struts," Haise said. "I solved their problem on the fifth landing flight, where I landed on the runway and bounced the vehicle, and my second landing was about five or six foot a second. So that gave them the data, and they were very happy with that—although I wasn't."

Haise gave a lot of the credit for the success of the ALT program to Deke Slayton, who agreed to run it after his return from his flight on the Apollo-Soyuz Test Project. "I frankly was very happy with Deke to volunteer for that role," he reflected.

I mean, we had no one, in my mind, that was at Johnson Space Center at the time that was better suited to take on that role. And I think it was reflected in the way the program went. We missed the first free-flight release from the 747 [by] only two weeks from a schedule that had been made several years before. We completed the program like four or five months earlier than we'd planned, which is almost unheard of in a test program, certainly something as complex as the orbiter. I think that was Deke's leadership in pulling together both the contingent of NASA, which involved a lot of integration of Kennedy Space Center people and Dryden NASA people, as well as the contractor Rockwell in that phase.

Although the ALT flights weren't spaceflights, they were still seen as plum assignments by many in the astronaut corps. Hank Hartsfield recalled being disappointed that he wasn't selected to one of the ALT crews. "I was extremely disappointed that I wasn't one of those, to be honest with you, and still don't know why. I mean, I thought that having developed the flight control system, I'd be in a good position. So did a lot of other people, but we learned along the way . . . crew assignments are strange things. You don't need to second-guess them. You just smile and press ahead."

Astronaut Joe Allen recalled an interesting political side of the *Enterprise* tests involving Senator Barry Goldwater.

Senator Goldwater had been a pilot in the Second World War, and he knew a lot about aviation. He was very proud of the aviation success of America. He sat on several NASA committees, was interested in NASA, and was a strong supporter of NASA, but he thought this was the most cockamamie idea he had ever seen, affixing the orbiter to the top of the 747 and then exploding away the bolts that held it there. He knew in his gut that, once released, the orbiter would slide back and hit the tail of the 747, break it off, and would be lost. He just knew it. He wanted hearings. I talked to his staff and said I would organize the hearing, but I requested fifteen minutes with the senator himself to go over the aerodynamics of the problem a little bit.

I brought a model and I said, "Senator Goldwater, I understand your concern, but I'm a pilot as well. Let me just talk as one pilot to another. No science here; we're just talking pilot talk."

Allen walked Goldwater through the process, step-by-step, explaining that because of the aerodynamics of the orbiter and the way it was mounted on the 747, its lift increased as the plane sped up, such that, on the ground, the plane was bearing the full weight of *Enterprise*, but toward the end of the flight, the orbiter was essentially weightless relative to the airplane. Shortly before the orbiter was released, the relationship changed so that *Enterprise* had enough lift that the orbiter was actually carrying some of the weight of the plane. Goldwater, he said, had no problem following when Allen broke the process down for him step-by-step.

"I said, 'Now let me show you the calculations. The tail drops, and by the time it goes below where the orbiter is, the orbiter has moved back only an eighth of an inch toward the tail, so it's not going to hit it.' And he said,

'I understand that. Why didn't NASA tell me that before?'"

No hearing was ever held.

Allen's political acumen came into play again in discussions with Indiana senator Birch Bayh, who sat on the Senate Appropriations Committee and whom Allen described as a rather liberal senator who was "not a friend of NASA."

"I'm from Indiana; I know Indiana people," Allen noted. "I got to know individuals in his office, including a most genuine lady who ran his office, and I discovered one day that she was from Rockville, Indiana. Rockville is a very tiny town. It has maybe two stoplights, or three, max. It is thirty miles from where I grew up in a somewhat bigger town, not much bigger. I knew something about Rockville that she did not know, and I got a photograph of the 747 with the *Enterprise* on top, flying along, a beautiful big photograph, and I took it in to this office."

When he delivered the picture, Bayh's secretary at first assumed it was for the senator. Allen corrected her.

I said, "No, this is for you. I brought this to you. I want to tell you something about this photo. This, of course, is the 747 and it's worth $300 million, and this is the orbiter, even more valuable." And I said, "The 747 on these tests is flown by an individual I think you know. His name is Tom McMurtry, and he grew up in Rockville, Indiana." He was a very skilled test pilot at NASA Dryden [Flight Research Center, Edwards, California]. And she said, "That's being flown by Tommy McMurtry?" I said, "Yes, that's correct." She said, "Golly. How much is all of that worth?" I said, "Well, it's about a billion and a half dollars." "Lordy," she says, "I remember when Tommy's daddy wouldn't let him drive the Buick." She was older than Tom, but she knew him as a boy. She immediately put the picture up on the senator's wall, and to my recollection the senator never voted against NASA again, ever, not once.

While the successful conclusion of the ALT flights brought the shuttle to launch, back in Florida, work to prepare *Columbia* for its maiden voyage was running into problems. In particular, the vital thermal protection tiles required to safely shield the vehicle from the heat of reentry were proving not to work as well in practice as they did in theory.

"Initially, when they put the tiles on," recalled Bob Crippen, *they weren't adhering to the vehicle like they should. In fact,* Columbia, *when they brought it from Edwards to the Kennedy Space Center the first time, it didn't arrive with as many tiles as it left California with. People started working very diligently to try to correct that problem, but at the same time people said, "Well, if we've got a tile missing off the bottom of the spacecraft that's critical to being able to come back in, we ought to have a way to repair it." So we started looking at various techniques, and I remember we took advantage of a simulator that Martin [Marietta Corporation] had out in Denver, where you could actually get some of the effect of crawling around on the bottom of the vehicle and what it would be like in zero g. I rapidly came to the conclusion I was going to tear up more tile than I could repair and that the only realistic answer was for us to make sure the tiles stayed on. . . . I concluded that at that time we couldn't have realistically repaired anything.*

Astronaut Charlie Bolden was part of Crippen's team in that effort. "Ideally, what we wanted was something that would be like a spray gun or something that you could just squirt and it would go into place, and then you could use a trowel and smooth it out, and you could fly home," he explained. "Everything seemed to be going very well initially, but every time we took whatever material was developed into vacuum, it just didn't work. . . . The gases in the material would just start to bubble out and cause it to crack and pop, and we became seriously concerned that the repair material would probably do as much damage or more than we had by a missing tile."

In the meantime, a new adhesive was developed that, it was believed, would greatly reduce the risk of tile loss. Additional analysis was also performed regarding the risks presented if there were tile loss in flight. Particular attention was paid to the possibility of a theoretical "zipper effect," in which the loss of one tile would cause aerodynamic forces to rip off more tiles and most likely doom the vehicle. The team eventually determined that the zipper effect was unlikely.

Bolden noted that, in all of the analysis of potential thermal protection risks, he does not recall any discussion about damage to the reusable carbon-carbon on the leading edge of the wing, which in 2003 resulted in the loss of *Columbia* and its crew.

Never in my memory—and I've been through my notebooks and everything—never did we talk about the reusable carbon-carbon, the RCC, *the leading edge of the wing, leading edge of the tail, and the nose cap itself. Nobody ever considered any damage to that because we all thought that it was impenetrable. In fact, it was not until the loss of* Columbia *that I learned how thin it was. I grew up in the space program. I spent fourteen years in the space program flying, thinking that I had this huge mass that was about five or six inches thick on the leading edge of the wing. And to find after* Columbia *that it was fractions of an inch thick, and that it wasn't as strong as the fiberglass on your Corvette, that was an eye-opener, I think, for all of us.*

Based on the success of the ALT program, NASA had looked to launch the first Space Shuttle mission by the end of 1979. However, a variety of issues, including those with the shuttle tiles, a problematic test firing of the Space Shuttle main engines, and an unexpected problem with insulation on the external tank after *Columbia* was stacked and on the launchpad, kept delaying the first launch of the Space Transportation System. While the delays were certainly not optimal, NASA and its astronauts worked to make the most of the extra time.

"Everybody wanted to get the bird in the air and show that it would fly okay, but the delays really provided us with more time to get ready for contingency situations," recalled Engle, who was in line to command the second shuttle mission. "I remember very distinctly not having the impression of idling or spinning our wheels or treading water during those delays. We were engrossed in always new things to look at. . . . But we were very, very busy getting ready for things, overpreparing, I think in retrospect."

According to Engle, the delays also helped the astronauts and Mission Control develop further into an effective team.

Because of the launch delays, we did have additional time to prepare for the flights and we got to work very closely and come to know and have a rapport with the controllers and, in fact, all the people in Mission Control, not just flight control, but all the people that were on the console, all [the] experts on the various systems. We knew them by voice, when we would hear transmissions. Of course, in the real flight, we only would talk to either the CapCom [capsule communicator, the astronaut in Mission Control who communicated with the crew], or sometimes pass information on to Don Puddy, the flight director. But we got to know pretty much who was on the console by what kind of response or direction we were given for certain simulated failures that we'd had during simulations. And that was good; that was really good. We worked as a very, very close-knit team, almost being able to think and read each other without a whole lot of words said.

Bob Crippen, the astronaut assigned to be pilot of the first shuttle launch, recalled touring NASA and its contractors during that period with STS-1 commander John Young to encourage the teams that were preparing their vehicle for flight.

The Space Shuttle is a pretty complicated vehicle, and certainly it was breaking some new frontiers, which, John and I being test pilots, it was a great mission for us. But when you have something that complicated, literally hundreds of thousands of people made STS-1 possible. You have to be depending on those folks doing their job right, because you can't check everything yourself. John and I spent a lot of time going to the various contractors and subcontractors, if nothing else, to try to put a human face on the mission; that we were flying it, and we appreciated all the work that they were doing.

Everywhere we went, the people really felt in their heart that they were doing something important for the nation, and that's what John and I wanted them to feel, because that's what they were doing. It doesn't take but one person to do something wrong that can cost you a mission and cost lives. So when John and I climbed aboard the vehicle to go fly, we had been eyeball to eyeball with, I would say, maybe not all hundred thousand, but we had been eyeball to eyeball with thousands of folks.

The visits not only served to encourage the workforce responsible for preparing the vehicle for flight, they gave the shuttle's first crew a unique, up-close look at what went into what the astronauts would be flying.

"We knew the vehicle pretty good," Crippen said.

We knew its risks, but we also knew that it had a great deal to offer if it was successful. So when we climbed aboard, I think both of us felt very confident that the vehicle would fly and fly well.

So when I say that we literally rode on the shoulders of hundreds of thousands of people, that's what we did, and I felt good about it. . . . I did have the opportunity after the first mission to go around again and thank a lot of the people that made the mission a success. Signed lots of autographs and did those kinds of things, to hopefully make them feel good about the work they had done. That continues today with the program of Spaceflight Awareness, because it's impor-

tant that they get feedback about how important their work is. . . . We're proud of what they're doing, and what they're doing is something great for the nation, and not only the nation, for the world, in my perspective. I'm proud of the shuttle. I'm proud to have been a part of it since almost its inception. . . . I think all the people that have been part of it ought to be proud of the job they've done.

5. First Flight

Well, we were just finally glad to get it started.
Geez, we'd waited so long. . . . We're finally getting
back into the flying business again.

—*Astronaut Owen Garriott*

STS-1
Crew: Commander John Young, Pilot Bob Crippen
Orbiter: *Columbia*
Launched: 12 April 1981
Landed: 14 April 1981
Mission: Test of orbiter systems

It was arguably the most complex piece of machinery ever made. And it had to work right the first time out. No practice round, just one first flight with two lives in the balance and the weight of the entire American space program on its shoulders.

The Redstone, Atlas, and Titan boosters used in the Mercury and Gemini programs had flight heritage prior to their use for manned NASA missions, but even so, the agency ran them through further test flights before putting humans atop them. And thankfully so, since their conversion from missiles into manned rockets was not without incident. There was the Mercury-Redstone vehicle that flew four inches on a test flight before settling back onto the pad. Or the Mercury-Atlas rocket that exploded during a test flight before the watching eyes of astronauts who were awaiting a ride on the launch vehicle in the future.

Unlike those used in Mercury and Gemini, the Saturn rockets used in the Apollo missions were newly designed by NASA for the purpose of manned spaceflight and had no prior history. When the first unmanned Saturn V was launched fully stacked for an "all-up" test flight in 1967, it was a major departure from the incremental testing used for the smaller Saturn I, in which the first stage was demonstrated to be reliable by itself before being flown with a live upper stage. The decision to skip the stage-by-stage testing for the Saturn V was a key factor in meeting President John F. Kennedy's "before this decade is out" goal for the first lunar landing, but it was not without a higher amount of risk.

That risk, however, paled next to the risk involved in the first flight of the Space Shuttle system. Not only could there be no incremental launches of the components, but there would be no unmanned test flight, no room for the sorts of incidents that occurred back in the Mercury testing days. The first Space Shuttle launch would carry two astronauts, and it had to carry them into space safely and bring them home successfully. Although it would be well over a decade before a movie scriptwriter would coin a phrase that would define the *Apollo 13* mission, it definitely applied in this case: failure was not an option.

Rather than being upset over the greater risk, however, NASA's astronauts endorsed the decision, some rather passionately. Most of the more veteran members of the corps had come to NASA as test pilots, and they were about to be test pilots again in a way they had never before experienced as astronauts. In fact, some pushed against testing the vehicle unmanned because of the precedent it would set.

"Oh, we were tickled silly," said Joe Engle, "because we didn't want any autopilot landing that vehicle."

Conducting an unmanned test flight would mean developing the capability for the orbiter to fly a completely automated mission, meaning that it would be capable of performing any of the necessary tasks from launch to landing. Some in the Astronaut Office were concerned that the more the orbiter was capable of doing by itself, the less NASA would want the astronauts to do when crewed flights began.

"We were very vain," Engle added,

and thought, you've got to have a pilot there to land it. If you've got an airplane, you've got to have a pilot in it. Fortunately, the certification of the autopilot all the way down to landing would have required a whole lot more cost and development time, delay in launch, and I think the rationale that we put forward

12. Space Shuttle *Columbia* poised for takeoff of STS-1. Courtesy NASA.

to discourage the idea of developing the autoland—and I think [a] correct one, too—was that you can leave it engaged down to a certain altitude, but always you have to be ready to assume that you're going to have an anomaly in the autopilot and the pilot has to take over and land anyway. The pilot flying the vehicle all the way down—approach and landing—he's in a much better position to affect the final landing, having become familiar and acclimated to the responses of the vehicle after being in orbit and knowing what kinds of displacements give him certain types of responses. Keeping himself lined up in the groove coming down to land is a much better situation than asking him to take over after autopilot has deviated off.

According to astronaut Hank Hartsfield, one of the major technical limitations that made the idea of an unmanned test flight particularly challenging was that the shuttle's computer system—current for the day but primitive by modern standards—would have to launch with all of the software needed for the entire mission already set up; there would be no crew to change out the software in orbit.

"We had separate software modules that we had to load once on orbit," Hartsfield explained. "All that had to be loaded off of storage devices or carried in core, because, remember, the shuttle computer was only a 65K machine. If you know much about storage, it's small, 65K. When you think about complexity of the vehicle that we're flying—a complex machine with full computers voting in tight sync and flying orbital dynamic flight phase and giving the crew displays and we're doing that with 65K of memory—it's mind-boggling."

Acknowledging the test-flight nature of the first launches, a special modification was made to *Columbia*—ejection seats for the commander's and pilot's seats on the flight deck. According to STS-1 pilot Bob Crippen, the ejection seats were a nice gesture but, in reality, probably would have made little difference in most scenarios.

"People felt like we needed some way to get out if something went wrong," Crippen said.

In truth, if you had to use them while the solids were there, I don't believe if you popped out and then went down through the fire trail that's behind the solids that you would have ever survived, or if you did, you wouldn't have a parachute, because it would have been burned up in the process. But by the time the solids had burned out, you were up to too high an altitude to use it.

On entry, if you were coming in short of the runway because something had happened, either you didn't have enough energy or whatever, you could have ejected. However, the scenario that would put you there is pretty unrealistic. So I personally didn't feel that the ejection seats were really going to help us out if we really ran into a contingency. I don't believe they would have done much for you, other than maybe give somebody a feeling that, hey, well, at least they had a possibility of getting out. So I was never very confident in them.

Risky or not, the two crew assignment slots that were available for the mission were a highly desired prize within the astronaut corps. STS-1 would be the first time any U.S. astronaut had flown into space in almost six years, would offer the opportunity to flight test an unflown vehicle, and would guarantee a place in the history books. Everyone wanted the flight. Senior astronaut John Young and rookie Bob Crippen got it.

Crippen recalled when he learned that he would be the pilot for the first Space Shuttle mission but said he still didn't know exactly why he was the one chosen for the job.

"Beats the heck out of me," he said.

I had anticipated that I would get to fly on one of the shuttle flights early on, because there weren't that many of us in the Astronaut Office during that period of time. I was working like everybody else was working in the office, and there were lots of qualified people. One day we had the Enterprise *coming through on the back of the 747. It landed out at Ellington[Field, Houston]. . . . I happened to go out there with George Abbey, who at that time was the director of flight crew operations. As we were strolling around the vehicle, looking at the* Enterprise *up there on the 747, George said something to the effect of, "Crip, would you like to fly the first one?"*

About that time I think I started doing handsprings on the tarmac out there. I couldn't believe it. It blew my mind that he'd let me go fly the first one with John Young, who was the most experienced guy we had in the office, obviously, and the chief of the Astronaut Office. It was a thrill. It was one of the high points of my life.

To be sure, Crippen's background made him an ideal candidate for the position. Crippen had been brought into NASA's astronaut corps in 1969 in a group that had been part of the air force's astronaut corps. After the cancellation of the air force's Manned Orbiting Laboratory program, NASA agreed to bring most of the air force astronauts who had been involved in the MOL program into the NASA astronaut corps.

Crippen's first major responsibility at NASA was supporting the Skylab space station program, followed by supporting the shuttle during its development phase. "I guess all that sort of added up, building on my experience. Working on the shuttle, I did primarily the software stuff, computers, which I enjoyed doing. And I think all that, stacked up together, kind of opened up the doors for me to fly. . . . I'm not sure whose decision that was, whether it was John Young's, George Abbey's, or who knows, but I'm sure glad they picked me."

Interestingly, Crippen had been part of the earliest wave of college students who had amazing opportunities open up for them by becoming savvy in the world of computer technology. During his senior year he took the first computer programming class ever offered at the University of Texas. "Computers were just starting to—shows you how old I am—to be widely used. . . . Not PCs or anything like that, but big mainframe kind of things, and Texas offered a course, and I decided I was interested in that and I

13. STS-1 crew members Commander John Young and Pilot Bob Crippen. Courtesy NASA.

would try it. It was fun. That was back when we were doing punch cards and all that kind of stuff."

With that educational background under his belt, Crippen took advantage of an opportunity with the air force Manned Orbiting Laboratory program to work on the computers for that vehicle. That in turn carried over, after his move to NASA, to working on computers for Skylab, which were similar to the ones for MOL.

"It was kind of natural, when we finished up Skylab and I started working on the Space Shuttle, to say, 'Hey, I'd like to work on the computers,'" Crippen said. "T. K. Mattingly was running all the shuttle operations in the Astronaut Office at that time and didn't have that many people that wanted to work on computers, so he said, 'Go ahead.' . . . The computer sort of interfaced with everything, so it gives you an opportunity to learn the entire vehicle."

Of course, being named pilot for STS-1 didn't mean that Crippen would actually pilot the vehicle. "We use the terms 'commander' and 'pilot' to confuse everybody, and it's really because none of us red-hot test pilots want to be called a copilot. In reality, the commander is the pilot, and the pilot is a copilot, kind of like a first officer if you're flying on a commercial airliner. [My job on STS-1] was primarily systems oriented, working the computers, working the electrical systems, working the auxiliary power units, doing the payload bay doors."

Crippen considered it an honor to work with the senior active member of the Astronaut Office for the flight. "When you're a rookie going on a test flight like this, you want to go with an old pro, and John was our old pro," he said.

He had four previous flights, including going to the moon, and John is not only an excellent pilot, he's an excellent engineer. I learned early on that if John was worried about something, I should be worried about it as well. This was primarily applying to things that we were looking into preflight. It's important for the commander to sort of set the tone for the rest of the crew as to what you ought to focus on, what you ought to worry about, and what you shouldn't worry about. I think that's the main thing I got out of John.

Crippen described Young as one of the funniest men he ever knew and regretted that he didn't keep record of all the funny things Young would say. "He's got a dry wit that a lot of people don't appreciate fully at first," Crippen said, "but he has got so many one-liners. If I had just kept a log of all of John's one-liners during those three years of training, I could have published a book, and he and I could have retired a long time ago. He really is a great guy."

The crew was ready; the vehicle less so. There were delays, followed by delays, followed by delays. Recalled astronaut Bo Bobko, "John Young had come up to me one day and said, 'Bo,' he said, 'I'd like you to take a group of guys and go down to the Cape and kind of help get everything ready down there.' And he said, 'I know it's probably going to take a couple of months.' Well, it took two years."

Much of that two years was waiting for things beyond his control, things taking place all over the country, Bobko said. However, he said, even the time spent waiting was productive time. He and others in the corps participated in the development and the testing that were going on at the Cape. Said Bobko,

Give you an example. I gave Dick Scobee the honor of powering up the shuttle the first time. I think it was an eight o'clock call to stations and a ten o'clock test start in the morning. I came in the next morning to relieve him at six, and he

14. Space Shuttle *Columbia* arrives at Launchpad 39A on 29 December 1980. Courtesy NASA.

wouldn't leave until he had thrown at least one switch. They had been discussing the procedures and writing deviations and all that sort of thing the whole day before and the whole night, and so they hadn't thrown one switch yet. So there was a big learning curve that we went through.

Finally, the ship was ready.

Launch day came on 10 April 1981. Came, and went, without a launch. The launch was scrubbed because of a synchronization error in the orbiter's computer systems. "The vehicle is so complicated, I fully anticipated that we would go through many, many countdowns before we ever got off," Crippen recalled. "When it came down to this particular computer problem, though, I was really surprised, because that was the area I was supposed to know, and I had never seen this happen; never heard of it happening."

Young and Crippen were strapped in the vehicle, lying on their backs, for a total of six hours. "We climbed out, and I said, 'Well, this is liable to take months to get corrected,' because I didn't know what it was. I'd never seen it," Crippen said. "It was so unusual and the software so critical to us. But we had, again, a number of people that were working very diligently on it."

The problem was being addressed in the Shuttle Avionics Integration Laboratory, which was used to test the flight software of every Space Shuttle flight before the software was actually loaded onto the vehicle. Astronaut Mike Hawley was one of many who were working on solving the problem and getting the first shuttle launch underway. "We at SAIL had the job of trying to replicate what had happened on the orbiter, and I was assigned to that," Hawley explained. "What happened when they activated the backup flight software, . . . it wasn't talking properly with the primary software, and everybody assumed initially that there was a problem with the backup software. But it wasn't. It was actually a problem with the primary software."

When the software activates, he explained, it captures the time reference from the Pulse Coded Modulation Master Unit. On rare occasions, that can happen in a way that results in a slightly different time base than the one that gets loaded with the backup software. If they don't have the same time base, they won't work together properly, and that was what had happened in this case.

So what we had to do was to load the computers over and over and over again to see if we could find out whether we could hit this timing window where the software would grab the wrong time and wouldn't talk to the backup. I don't remember the number. It was like 150 times we did it, and then we finally recreated the problem, and we were able to confirm that. So that was nice, because that meant the fix was all you had to do was reload it and just bring it up again. It's now like doing the old control-alt-delete on your PC. It reboots and then everything's okay. And that's what they did.

Crippen said SAIL concluded that the odds of that same problem occurring again were very slim. This was all sorted out in just one day. "We scrubbed on the tenth, pretty much figured out the problem on the eleventh, and elected to go again on the twelfth."

Crippen, though, had little faith that the launch would really go on the twelfth. "I thought, 'Hey, we'll go out. Something else will go wrong. So we're going to get lots of exercises at climbing in and out.' But I was wrong again."

On 12 April 1981 the crew boarded the vehicle, which in photos looks distinctive today for its white external tank. The white paint was intended to protect the tank from ultraviolet radiation, but it was later determined that the benefits were negligible and the paint was just adding extra weight.

After the second shuttle flight, the paint was no longer used. Doing without it improved the shuttle's payload capacity by six hundred pounds.

Strapped into the vehicle, the crew began waiting. And then waiting a little bit more. "George Page, a great friend, one of the best launch directors Kennedy Space Center has ever had, really ran a tight control room," Crippen recalled. "He didn't allow a bunch of talking going on. He wanted people to focus on their job. In talking prior to going out there, George told John and I, he says, 'Hey, I want to make sure everybody's really doing the right thing and focused going into flight. So I may end up putting a hold in that is not required, but just to get everybody calmed down and making sure that they're focused.' It turned out that he did that."

Even after that hold, as the clock continued to tick down, Crippen said, he still was convinced that a problem was going to arise that would delay the launch.

It's after you pass that point that things really start to come up in the vehicle, and you are looking at more systems, and I said, "Hey, we're going to find something that's going to cause us to scrub again." So I wasn't very confident that we were going to go. But we hadn't run into any problem up to that point, and when . . . I started up the APUs, the auxiliary power units, everything was going good. The weather was looking good. About one minute to go, I turned to John. I said, "I think we might really do it," and about that time, my heart rate started to go up. . . . We were being recorded, and it was up to about 130. John's was down about 90. He said he was just too old for his to go any faster. And sure enough, the count came on down, and the main engines started. The solid rockets went off, and away we went.

Astronaut Loren Shriver had helped strap the crew members into their seats for launch and then left the pad to find a place to watch. He viewed the event from a roadblock three miles away, the closest point anyone was allowed to be for the launch, where fire trucks waited in case they were needed. "I remember when the thing lifted off, there were a number of things about that first liftoff that truly amazed me," he said.

One was just the magnitude of steam and clouds, vapor that was being produced by the main engines, the exhaust hitting the sound suppression water, and then the solid rocket boosters were just something else, of course. And then when the sound finally hits you from three miles away, it's just mind-boggling. Even

15. The launch of STS-1 on 12 April 1981. Courtesy NASA.

for an experienced fighter pilot, test pilot, it was just amazing to stand through that, because being that close and being on top of the fire truck . . . the pressure waves are basically unattenuated except by that three miles of distance. But when it hits your chest, and it was flapping the flight suit against my leg, and it was vibrating, you could feel your legs and your knees buckling a little bit, could feel it in your chest, and I said, "Hmm, this is pretty powerful stuff here."

Crippen had trained with Young for three years, during which time the Gemini and Apollo astronaut imparted to Crippen as best he could what the launch would be like. "He told me about riding on the Gemini and riding on the [Apollo] command module—which he'd done two of each of those—and gave me some sense of what ascent was going to be like, what main engine cutoff was going to be like," Crippen recalled.

But the shuttle is different than those vehicles. I know when the main engines lit off, it was obvious that they had in the cockpit, not only from the instruments, but you could hear and essentially feel the vehicle start to shake a little bit. When the solids light, there was no doubt we were headed someplace; we were just hoping it was in the right direction. It's a nice kick in the pants; not violent. The thing that I have likened it to, being a naval aviator, is it's similar to a catapult shot coming off an aircraft carrier. You really get up and scoot, coming off the pad.

As the vehicle accelerated, the two-man crew felt vibration in the cockpit. "I've likened it to driving my pickup down an old country washboard road," explained Crippen.

It was that kind of shaking, but nothing too dramatic, and it . . . didn't feel as significant as what I'd heard John talk about on the Saturn V. When we got to two minutes into flight, when the SRBs came off, that was enlightening, . . . actually you could see the fire [from the separation motors] come over the forward windows. I didn't know that I was going to see that, but it was there, and it actually put a thin coat across the windows that sort of obscured the view a little bit, but not bad. But the main thing, when they came off, it had been noisy. We had been experiencing three gs, or three times the weight of gravity, and all of a sudden it was almost like there was no acceleration, and it got very quiet. . . . I thought for sure all the engines had quit. Rapidly checked my instruments, and they said no, we were still going. It was a big, dramatic thing, for me, at least. . . . You're up above most of the atmosphere, and it was just a very dramatic thing for me. It will always stick out [in my memory].

The engines throttled, like they were supposed to, and then, at eight and a half minutes after launch, it was time for MECO, or main engine cutoff. Crippen described the first launch of the nation's new Space Transportation System as quite a ride. "Eight and a half minutes from sitting on the pad to going 17,500 miles an hour is a ride like no other. It was a great experience."

Many astronauts over the years have commented just how much an actual shuttle launch is like the countless simulations they go through to prepare for the real thing. They have Young and Crippen to thank, in part, for the verisimilitude of the sims. Before they flew, there had been no "real thing" version on which to base the training, and when they returned, they brought back new insight to add to the fairly realistic "best-guess" version they had trained on. "I [went] back to the simulations," Crippen said.

We did change the motion-based, . . . to make it seem like a little bit more of a kick when you lifted off. We did make the separation boosters coming over the windscreen; we put that in the visual so that was there. We changed the shaking on that first stage somewhat so that it was at least more true to what the real flight was like. . . . After we got the main engine cutoff and the reaction control jets started firing, we changed that noise to make sure it got everybody's attention so that they wouldn't be surprised by that.

With the main engine cutoff, *Columbia* had made it successfully into orbit. The duo was strapped so tightly into their seats they didn't feel the effects of microgravity right away. But then the checklists started to float around, and even though the ground crews had worked very hard to deliver a pristinely clean cockpit, small debris started to float around too.

Now the next phase of the mission began. On future missions, nominal operations of the spacecraft would be almost taken for granted during the orbital phase of the mission. On STS-1, to a very real extent, the nominal operations of the spacecraft were the focus of the mission. "On those initial flights, including the first one, we only had two people on board, and there was a lot to do," Crippen said.

We didn't have any payloads, except for instrumentation to look at all of the vehicle. So we were primarily going through what I would call nominal things for a flight, but they were being done for the first time, which is the way a test flight would be done. First you want to make sure that the solids would do their thing, that the main engines would run, and that the tank would come off properly, and that you could light off the orbital maneuvering engines as planned; that the payload bay doors would function properly; that you could align the inertial measuring units; the star trackers would work; the environmental control system, the Freon loops, would all function. So John and I, we were pretty busy. The old "one-armed paper hanger" thing is appropriate in this case. But we did find a little time to look out the windows, too.

For Young, it was his fifth launch into space—or technically, sixth, counting his space launch from the surface of the moon during *Apollo 16*. For Crippen, it was his first, and the experience of being in the weightlessness of orbit was a new one. "We knew people had a potential for space sickness, because that had occurred earlier, and the docs made me take some medication before liftoff just in case," he said. "I was very sensitive when we got on orbit as to how I would move around. I didn't want to move my head too fast. I didn't want to get flipped upside down in the cockpit. So I was moving, I guess, very slowly."

Moving deliberately, Crippen eased to the rear of the flight deck to open up the payload bay doors. Other than the weightlessness, Crippen said, the work he was doing seemed very much like it had during training. "I said, 'You know, this feels like every time I've done it in the simulator, except my feet aren't on the floor.' . . . The simulations were very good. So I went ahead and did the procedure on the doors. Unlatched the latches; that worked great. Opened up the first door, and at that time I saw, back on the Orbital Maneuvering System's pods that hold those engines, that there were some squares back there where obviously the tiles were gone. They were dark instead of being white."

The tiles, of course, were part of the shuttle's thermal protection system, which buffers it against the potentially lethal heat of reentry. While the discovery that some tiles were missing was an obvious source of concern for some, Crippen said he wasn't among them.

I went ahead and completed opening the doors, and when we got ground contact . . . we told the ground, "Hey, there's some tile missing back there," and we gave them some TV views of the tiles that were missing. Personally, that didn't cause me any great concern, because I knew that all the critical tiles, the ones primarily on the bottom, we'd gone through and done a pull test with a little device to make sure that they were snugly adhered to the vehicle. Some of them we hadn't done, and that included the ones back on the OMS pods, and we didn't do them because those were primarily there for reusability, and the worst that would probably happen was we'd get a little heat damage back there from it.

The concern on the ground, however, dealt with what else the loss of those tiles might mean. If thermal protection had come off of the OMS pods, could tiles have also been lost in a more dangerous area—the underside of the orbiter, which the crew couldn't see? Crippen said that he and Young chose not to worry about the issue since, if there was a problem, there was nothing they could do about it at that point.

As the "capsule communicator," or CapCom (a name that dates back to the capsules used for previous NASA manned missions), for the flight, astronaut Terry Hart was the liaison in Mission Control between the crew and the flight director and so was in the middle of the discussions about the tile situation. "Here were some tiles missing on the top of the OMS pods, the engine pods in the back, which immediately raised a concern," Hart said.

Was there something underneath missing, too? Of course, we'd had all these problems during the preparation, with the tiles coming off during ferry flights and so forth, and the concern was real. I think they found some pieces of tiles in the flame trench [under the launchpad] after launch as well, so there was kind of a tone of concern at the time, not knowing what kind of condition the bottom of the shuttle was in, and we had no way to do an inspection. . . . All of a sudden the word kind of started buzzing around Mission Control that we don't have to worry anymore. So we all said, "Why don't we have to worry anymore?" "Can't tell you. You don't have to worry anymore." So about an hour later, Gene Kranz walked in and he had these pictures of the bottom of the shuttle. It was, "How did you get those?" He said, "I can't tell you." But we could see that the shuttle was fine, so then we all relaxed a little bit and knew that it was going to come back just fine, which it did.

Hart said that the mystery was revealed for him many years later, after the loss of *Columbia* in 2003, explaining that it came out that national defense satellites were able to image the shuttles in orbit.

Despite Crippen's concerns about space sickness that caused him to be more deliberate during the door-opening procedure, he said that the issue proved not to be a problem for him. "I was worried about potentially being sick, and it came time after we did the doors to go get out of our launch escape suits, the big garments we were wearing for launch. I went first and went down to the mid-deck of the vehicle and started to unzip and climb out of it, and I was tumbling every which way and slipped out of my suit and concluded, 'Hey, if I just went and tumbled this way and tumbled that way, and my tummy still felt good, then I didn't have to worry about getting sick, thank goodness.'"

In fact, Crippen found weightlessness to be rather agreeable.

This vehicle was big enough, not like the [Apollo] command modules or the Gemini or Mercury, that you could move around quite a bit. Not as big as Skylab, but you could take advantage of being weightless, and it was delightful. It was a truly unique experience, learning to move around. I found out that it's always good to take your boots off—which I had taken mine off when I came out of the seat—

because people, when they get out and then being weightless for the first time, they tend to flail their feet a little bit like they were trying to swim or something. So I made sure on all my crews after that, they know that no boots, no kicking.

A few of the issues encountered on *Columbia*'s maiden voyage were problems with the bathroom and trying several times to access a panel that, unbeknownst to the crew, had been essentially glued shut by one of the ground crew. However, there were plenty of things that went well. One of those, Crippen said, was the food, which was derived from the food used during Skylab missions. "We even had steak. It was irradiated so that you could set it on the shelf for a couple of years, open it up, and it was just like it had come off the grill. It was great. We had great food, from my perspective."

STS-1 was, in and of itself, a historic event, but another historic event around the same time would have an impact on a key moment of the flight, Crippen recalled.

There was an established tradition of the president talking to astronauts on historic spaceflights, and the first flight of the new vehicle and the first American manned spaceflight in almost six years ordinarily would have been no exception. But, in that regard, the timing of STS-1 was anything but ordinary. President Reagan was shot two weeks prior to our flight, so . . . the vice president called, as opposed to the president. The vice president, [George H. W.] Bush, had also come to visit us at the Cape. It was sometime prior to flight, but we had him up in the cockpit of Columbia *and looked around, went out jogging a few miles with him. So we felt like we had a personal rapport with him, and so when we got a call from the vice president, it was like talking to an old friend.*

The launch had been successful; the orbital phase had been successful, demonstrating that the vehicle functioned properly while in space; and now one more major thing remained to be proven: that what had gone up could safely come back down. "That was also one of these test objectives, to make sure that we could deorbit properly," Crippen said.

We did our deorbit burn on the dark side of the Earth and started falling into the Earth's atmosphere. It was still dark when we started to pick up outside the window; it turned this pretty color of pink. It wasn't a big fiery kind of a thing like they had with the command modules; . . . they used the ablative heat shield. It was just a bunch of little angry ions out there that were proving that it was kind of warm outside, on the order of three thousand degrees out the front window. But it was pretty. It was kind of like you were flying down a neon tube, about that color of pink that you might see in a neon tube.

At that point in the reentry, the autopilot was on and things were going well, according to Crippen. He said that Young had been concerned with how the vehicle would handle the S-turns deeper into the atmosphere and took over from the autopilot when the orbiter had slowed to about Mach 7, the first of a few times he switched back and forth between autopilot and manual control, until taking over manually for good at Mach 1. At the time of the first switch, the crew was out of contact with Mission Control. "We had a good period there where we couldn't talk to the ground because there were no ground stations," Crippen explained. "So I think the ground was pretty happy the first time we reported in to them that we were still there, coming down."

Astronaut Hank Hartsfield recalled those moments of silence from the perspective of someone on the ground. "It was an exciting time for me when we flew that first flight, watching that and seeing whether we was going to get it back or not," he said. "Of course, we didn't have comm all the way, and so once they went LOS [loss of signal] in entry was a real nervous time to see if somebody was going to talk to you on the other side. It was really great when John greeted us over the radio when they came out of the blackout."

Continued Crippen,

I deployed the air data probes around Mach 5, and we started to pick up air data. We started to pick up TACANs [Tactical Air Control and Navigation systems] to use to update our navigation system, and we could see the coast of California. We came in over the San Joaquin Valley, which I'd flown over many times, and I could see Tehachapi, which is the little pass between San Joaquin and the Mojave Desert. You could see Edwards, and you could look out and see Three Sisters, which are three dry lake beds out there. It was just like I was coming home. I'd been there lots of times. I did remark over the radio that, "What a way to come to California," because it was a bit different than all of my previous trips there.

Young flew the shuttle over Edwards Air Force Base and started to line up for landing on Lake Bed 22. Crippen recalled,

My primary job was to get the landing gear down, which I did, and John did a beautiful job of touching down. The vehicle had more lift or less drag

16. The Space Shuttle *Columbia* glides in for landing at the conclusion of STS-1. Courtesy NASA.

than we had predicted, so we floated for a longer period than what we'd expected, which was one of the reasons we were using the lake bed. But John greased it on.

Jon McBride was our chase pilot in the T-38. I remember him saying, "Welcome home, Skipper," talking to John. After we touched down, John was . . . feeling good. Joe Allen was the CapCom at the time, and John said, "Want me to take it up into the hangar, Joe?" Because it was rolling nice. He wasn't using the brakes very much. Then we got stopped. You hardly ever see John excited. He has such a calm demeanor. But he was excited in the cockpit.

On the ground, the crew still had a number of tasks to complete and was to stay in the orbiter cockpit until the support crew was on board. "John unstrapped, climbed down the ladder to the mid-deck, climbed back up again, climbed back down again," recalled Crippen.

He couldn't sit still, and I thought he was going to open up the hatch before the ground did, and of course, they wanted to go around and sniff the vehicle and make sure that there weren't any bad fumes around there so you wouldn't inhale them. But they finally opened up the hatch, and John popped out. Meanwhile I'm still up there, doing my job, but I will never forget how excited John was. I completed my task and went out and joined him awhile later, but he was that excited all the way home on the flight to Houston, too.

Crippen's additional work after Young left the orbiter meant that the pilot had the vehicle to himself for a brief time, and he said that those postflight moments alone with *Columbia* were another memory that he would always treasure.

CapCom Joe Allen recalled the reaction in Mission Control to the successful completion of the mission.

When the wheels stopped, I was very excited and very relieved. . . . I do remember that Donald R. Puddy, who was the entry flight director, said, "All controllers, you have fifteen seconds for unmitigated jubilation, and then let's get this flight vehicle safe," because we had a lot of systems to turn down. So people yelled and cheered for fifteen seconds, and then he called, "Time's up." Very typical Don Puddy. No nonsense. . . . Other images that come into my head, the people there that went out in the vans to meet the orbiter wear very strange looking protective garments to keep nasty propellants that a Space Shuttle could be leaking from harming the individuals. These ground crew technicians look more like astronauts than the astronauts themselves. Then the astronauts step out and in those days, they were wearing normal blue flight suits. They looked like people, but the technicians around them looked like the astronauts, which I always thought was rather amusing.

Terry Hart was CapCom during the launch of STS-1, but he said watching the landing overshadowed his launch experience. The landing was on his day off and he had no official duties, yet he watched from the communications desk inside Mission Control. "When the shuttle came down to land, I had tears in my eyes," said Hart. "It was just so emotional. On launch, typically you're just focused on what you have to do and everything, but to watch the *Columbia* come in and land like that, it was really beautiful and it was kind of like a highlight. Even though I wasn't really involved, I could actually enjoy the moment more by being a spectator."

Astronaut John Fabian recalled thinking during the landing of that first mission just how risky it was.

The risk of launch is going to be there regardless of what vehicle you're launching, and the shuttle has some unique problems; there's no question about that. But this is the first time; we've never flown this thing back into the atmosphere. We don't have any end-to-end test on those tiles. The guidance system has only been simulated; it's never really flown a reentry. And this is really hairy stuff that's going on out there. And the feeling of elation when that thing came back in and looked good—the landing I never worried about, these guys practice a thousand landings, but the heat of reentry, it was something that really, really was dangerous.

While the mission was technically over, the duties of the now-famous first crew of the Space Shuttle were far from complete. The two would continue to circle the world, albeit much more slowly and closer to the ground, in a series of public relations trips. "The PR that followed the flight I think was somewhat overwhelming, at least for me," Crippen said. "John was maybe used to some of it, since he had been through the previous flights. But we went everywhere. We did everything. That sort of comes with the territory. I don't think most of the astronauts sign up for the fame aspect of it."

In particular, he recalled attending a conference hosted by ABC television in Los Angeles very shortly after the flight.

All of a sudden, they did this grandiose announcement, and you would have thought that a couple of big heroes or something were walking out. They were showing all this stuff, and they introduced John. We walk out, and there's two thousand people out there, and they're all standing up and applauding. It was overwhelming.

We got to go see a lot of places around the world. Did Europe. Got to go to Australia; neat place. In fact, I had sort of cheated on that. I kept a tape of "Waltzing Matilda" and played it as we were coming over one of the Australian ground stations, just hoping that maybe somebody would give us an invite to go to Australia, since I'd never been there. But that was fun.

Prior to STS-1, Crippen recalled, the country's morale was not very high. "We'd essentially lost the Vietnam War. We had the hostages held in Iran. The president had just been shot. I think people were wondering whether we could do anything right. [STS-1] was truly a morale booster for the United States. . . . It was obvious that it was a big deal. It was a big deal to the military in the United States, because we planned to use the vehicle to fly military payloads. It was something that was important."

Crippen said STS-1 and the nation's return to human spaceflight provided a positive rallying point for the American people at the time, and human space exploration continues to have that effect for many today. "A great many of the people in the United States still believe in the space program," he said. "Some think it's too expensive. Perspective-wise, it's not that expensive, but I believe that most of the people that have come in contact with the space program come away with a very positive feeling. Sometimes, if they have only seen it on TV, maybe they don't really understand it, and there are some negative vibes out there from some individuals, but most people, certainly the majority, I think, think that we're doing something right, and it's something that we should be doing, something that's for the future, something that's for the future of the United States and mankind."

6. The Demonstration Flights

With the return of *Columbia* and its crew at the end of STS-1, the first flight was a success, but the shuttle demonstration flight-test program was still just getting started. Even before STS-1 had launched, preparations for STS-2 were already under way. The mission would build on the success of the first and test additional orbiter systems.

STS-2
Crew: Commander Joe Engle, Pilot Dick Truly
Orbiter: *Columbia*
Launched: 12 November 1981
Landed: 14 November 1981
Mission: Test of orbiter systems

STS-1 had been a grand experiment, the first time a new NASA human launch vehicle made its maiden flight with a crew aboard. Because of that, the mission profile was kept relatively simple and straightforward to decrease risk. Before the shuttle could become operational, however, many more of its capabilities would have to be tested and demonstrated. In that respect, the mission of STS-2 began long before launch.

"One of the things I remember back then on STS-2," recalled astronaut Mike Hawley,

when they mate the shuttle to the solid rocket motor on the external tank in the Vehicle Assembly Building [VAB, at Kennedy Space Center in Florida], they do something called the shuttle interface test, . . . making sure the cables are hooked up right and the hydraulic lines are hooked up properly and the orbiter and the solids and all of that functions as a unit before they take it to the pad. In those days we manned that test with astronauts, and for STS-2 the astronauts that manned it were me and Ellison Onizuka. We were in the Columbia *in the middle of the night hanging on the side of the ET [external tank] in the VAB going through this test.*

Part of the test, Hawley explained, involved making sure that the orbiter's flight control systems were working properly, including the moving flight control surfaces, the displays, and the software.

It turns out when you do part of this test and you bypass a surface in the flight control system, it shakes the vehicle and there's this big bang that happens. Well, nobody told me that, and Ellison and I are sitting in the vehicle going through this test, and I forget which one of us threw the switch, but . . . there's this "bang!" and the whole vehicle shakes. We're going, "Uh-oh, I think we broke it." But that was actually normal. I took delight years later in knowing that was going to happen and not telling other people, so that they would have the same fun of experiencing what it's like to think you broke the orbiter. . . . That was a lot of fun. Except for flying, that was probably the most fun I ever had, was working the Cape job.

In contrast to STS-1, commanded by NASA's senior member of the astronaut corps, STS-2 was the only demonstration flight to be commanded by a "rookie" astronaut—Joe Engle, who had earned air force astronaut wings on a suborbital spaceflight on the X-15 rocket plane.

Engle's path to becoming the shuttle's first NASA-unflown commander began about a decade earlier, when he lost out on a chance to walk on the moon. Engle had been assigned to the crew of *Apollo 17*, alongside commander Gene Cernan and command module pilot Ron Evans. However, pressure to make sure that geologist-astronaut Harrison Schmitt made it to the moon cost Engle his slot as the *Apollo 17* Lunar Module pilot. When Engle was informed that he was being removed from the mission, Deke Slayton discussed with him options for his next assignment.

"I wouldn't say I was able to select, but Deke was very, very good about it and asked me, in a one-to-one conversation, not promising that I would be assigned to a Skylab or not promising I would be assigned to the Apollo-Soyuz mission, but implying that if I were interested in that, he would sure consider that very heavily," Engle recalled.

He also indicated that the Space Shuttle looked like it was going to be funded and looked like it was going to be a real program. At the time, I think I responded something to the effect that it had a stick and rudder and wings and was an airplane and was kind of more the kind of vehicle that I felt I could contribute more to. And Deke concurred with that. He said, "That was my opinion, but if you want to fly sooner than that, I was ready to help out." I think Deke was happy that I had indicated I would just as soon wait for the Space Shuttle and be part of the early testing on that program.

The decision meant that Engle was involved in the very beginnings of the shuttle program, even before the contractors and final design for the vehicle had been selected.

Rockwell and McDonnell Douglas and Grumman were three of the primary competitors initially and had different design concepts for the shuttle, all pretty much the same, but they were significantly different in shape and in configuration for launch, using different types of boosters and things. I was part of that selection process only as an engineer and a pilot and assessing a very small part of the data that went into the final selection. But it was interesting. It was very, very interesting, and it, fortunately, allowed me to pull on some of the experience that I had gotten at Edwards in flight-testing, in trying to assess what might be the most reasonable approach to either flying initial flights, data-gathering, and things of that nature.

Engle and STS-2 pilot Richard Truly served as the backup crew for STS-1 and trained for that mission alongside Crippen and Young. Not long after the first mission was complete, a ceremony was held in which Engle and Truly were presented with a "key"—made of cardboard—to *Columbia.* "I think the hope was that would be a traditional handover of the vehicle to the next crew," Engle recalled.

It was a fun thing to do. In fact, I think it was done at a pilots' meeting one time, as I recall. John and Crip handed Dick and I the key, and I think there were so many comments about buying a used car from Crip and John that it became more of a joke than a serious tradition, and I don't recall that it really lasted very long. I think it turned from cardboard into plywood, and I don't recall that it was done very long after that. Plus, once we got the next vehicles, Discovery, Challenger, *on line, it lost some significance as well. You weren't really sure which vehicle you were going to fly after that, so you didn't know who to give the key to.*

In the wake of the tile loss on the first shuttle mission, a renewed effort was made to develop contingencies to deal with future problems and, according to Engle, additional progress was made.

Probably the biggest change that occurred was more emphasis on being able to make a repair for a tile that might have come off during flight. John and Crip lost a number of tiles on STS-1. Fortunately, none were in the critical underside, where the maximum heat is. Most of them were on the OMS pod and on the top of the vehicle. But the inherent cause of those tiles coming loose and separating was not really understood, and on STS-2 we were prepared to at least try to fill some of those voids with RSI [Reusable Surface Insulation], the rubbery material that bonds the tile to the surface itself. So in our training, we began to fold in EVA training, using materials and tools to fill in those voids.

Exactly seven months after the launch of STS-1, the launch date for the second Space Shuttle mission arrived. Sitting on the launchpad, Engle was in a distinctive position—on his first NASA launch, he would be returning to space. It would, however, be a very different experience from his last; he was actually going to stay in orbit versus just briefly skimming space, as he had in the X-15. In fact, he said, while his thoughts on the launchpad did briefly touch on his previous flights on the X-15, space wasn't the place he was thinking about returning to. Instead, he was already thinking ahead to the end of the mission.

As I recall, the only conscious recollection to the X-15 was that at the end of the flight we would be going back to Edwards and landing on the dry lake bed. And I think, for all of the training and all of the good simulation that we received, that's where I felt the most comfortable, the most at home, going back to Edwards. And at the end of the flight, when we rolled out on final approach going into the dry lake bed, that turned out to really be the case. It was a demanding mission and there were a lot of strange things that went on during our first flight, but when we got back into the landing pattern, it just felt like I was back at Edwards again, ready to land another airplane.

According to Engle, the launch was very much like it had been in the simulators. However, even though the first shuttle mission had provided real-world data on what an actual flight would be like, there were still, at the time the STS-2 crew was training, ways in which the simulators still did not fully capture the details of the ascent. The thing that most surprised him during launch was "the loud explosion and fireball when the solid rocket boosters were ejected. That was not really simulated very well in the sim-

17. An aerial view of the launch of *Columbia* on the second Space Shuttle mission, STS-2. Photo taken by astronaut John Young aboard NASA's Shuttle Training Aircraft. Courtesy NASA.

ulator, because I don't think anybody really anticipated it would be quite as impressive a show as that. I don't remember that [John and Crip] mentioned it to us, but that caught our attention, and I think we did pass that on in briefings to the rest of the troops, not just to [the STS-3 crew], but to everyone else who was flying downstream."

If the launch was relatively nominal, things changed quickly after that. Two and a half hours into flight, *Columbia* had a fuel cell failure, which, under mission rules, required an early return to Earth—cutting the mission from 125 hours to 54. "We were disappointed," Engle said.

As I recall, we kind of tried to hint that we probably didn't need to come back, we still had two fuel cells going, but at the time, it was the correct decision, be-

cause there was no really depth of knowledge as to why that fuel cell failed, and there was no way of telling that it was not a generic failure, that the other two might follow. And, of course, without fuel cells, without electricity, the vehicle is not controllable. So we understood and we accepted. We knew the ground rules; we knew the flight rules that dictated that if you lost a fuel cell, that it would be a minimum-time mission. We had really prepared and trained hard and had a full scenario of objectives that we wanted to complete on the mission. Of course, everybody wants to complete everything.

In fact, according to Engle, the only real disappointment he and Truly felt at the time was that, after investing so much training time into preparing for the mission goals, they weren't going to have the time to fully accomplish them. "I don't think we consciously thought, 'Well, we're not going to have five days to look at the Earth.' I don't think that really entered our minds right then, because we were more focused on how we are going to get all this stuff done."

Rather than accept defeat, Engle and Truly managed to get enough work done even in the shortened duration that the mission objectives were declared 90 percent complete at the end of the flight. "We were able to do it because we had trained enough to know precisely what all had to be done, and we prioritized things as much as we could," Engle explained.

Fortunately, we didn't have TDRSS [Tracking and Data Relay Satellite System] at the time. We only had the ground stations, so we didn't have continuous voice communication with Mission Control, and Mission Control didn't have continuous data downlink from the vehicle either, only when we'd fly over the ground stations. So when our sleep cycle was approaching, we did, in fact, power down some of the systems and we did tell Mission Control goodnight, but as soon as we went LOS, loss of signal, from the ground station, then we got busy and scrambled and cranked up the remote manipulator arm and ran through the sequence of tests for the arm, ran through as much of the other data that we could, got as much done as we could during the night. We didn't sleep that night; we stayed up all night. Then the next morning, when the wakeup call came from the ground, why, we tried to pretend like we were sleepy and just waking up.

After the flight, I remember Don Puddy saying, "Well, we knew you guys were awake, because when you'd pass over the ground station, we could see you were drawing more power than you should have been if you were asleep." But that was about the only insight they had into it.

The first use of the robot arm was one of the major ways that STS-2 moved forward from STS-1 in testing out the vehicle. "From the beginning we had the RMS, the remote manipulator system, the arm, manifested on our flight, and that was a major test article and test procedure to perform, to actually take the arm, to de-berth the arm and take it out through maneuvers and attach it to different places in the payload to demonstrate that it would work in zero gravity and work throughout its envelope. Dick became the primary RMS-responsible crew member and did a magnificent job in working with the arm people."

The premission training for testing the arm included working with its camera and figuring out what sort of angles it could capture. Truly learned that the arm could be maneuvered so that the camera pointed through the cabin windows of the orbiter. Once they were in orbit, Truly took advantage of this discovery by getting shots of himself and Engle in the cabin, with Truly displaying a sign reading "Hi, Mom."

Sally Ride, who was heavily involved in the training for the arm, also worked with the crew to help develop a plan for Earth observations during the mission. "They were both very, very interested in observing Earth while they were in space, but they weren't carrying instruments other than their eyes and their cameras," Ride said.

They wanted to have a good understanding of what they would be seeing, what they should look for, and what scientists wanted them to look for. The crew had to learn how to recognize the things that scientists were interested in: wave patterns on the surface of the ocean, rift valleys, large volcanoes, a variety of different geological features on the ground. I spent quite a bit of time with the scientists, as a liaison between the scientists and the astronauts who were going to be taking the pictures, to try to understand what the scientists had observed and then to help the astronauts understand how to recognize features of interest and what sort of pictures to take.

During the flight, not only was the crew staying extra busy to make up for the lost time, but CapCom Terry Hart recalled an incident in which the ground unintentionally made additional, unnecessary work for the crew.

It turned out that President Reagan was visiting Mission Control during the STS-2, and it was just [over eight months] after he had been shot. . . . This was one of his first public events since recovering from that. Of course, everyone was very excited about that, and it turned out he was coming in on our shift. . . . Mission Control was kind of all excited about the president coming and everything. He came in, and I was just amazed how large a man he was. I guess TV *doesn't make people look as large as they sometimes are. And I was on comm, so I was actually the one talking to the crew at that time when he came in. So I had the chance then to give him my seat and show him how to use the radios, and then I actually introduced him to the crew. I said something to the effect that, "*Columbia*, this is Houston. We have a visiting CapCom here today." I said, "I guess you could call him CapCom One." And then the president smiled and he sat down and had a nice conversation for a few minutes with the crew.*

While the event was a success and an exciting opportunity for the astronauts, Mission Control, and the president, Hart recalled that it was later discovered that a miscommunication had caused an unintended downside to the uplink.

We didn't realize we had not communicated properly to the crew, and the crew thought this was going to be a video downlink opportunity for them. . . . They had set up TV *cameras inside the shuttle to show themselves to Mission Control while they were talking to the president, when the plan was only to have an audio call with the president, because the particular ground station they were over at that time didn't have video downlink. We had wasted about an hour or two of the crew's time, so we kind of felt bad about that.*

We didn't learn about that until after. . . . It all worked out and everything, but it just shows you how important it is to communicate effectively with the crews and to work together as a team and all. And of course, most of the crews, the astronauts, they're all troopers. They want to do their very best, and if Mission Control is not careful, you'll let them overwork themselves.

The extra time the crew members gained by not sleeping at night allowed them to be more productive, but the lost sleep, and a side-effect problem related to the failed fuel cell, caught up with them as the mission neared its end. The membrane that failed on the fuel cell allowed excess hydrogen to get into the supply of drinking water. "So we had very bubbly water," Truly said.

Whenever we'd go to take a drink, . . . a large percentage of the volume was hydrogen bubbles in the water, and they didn't float to the top like bubbles would in a glass here and get rid of themselves, because in zero gravity they don't; they just stay in solution. We had no way to separate those out, so the water that we

18. From the Mission Control room at Johnson Space Center, President Ronald Reagan talks to Joe Engle and Richard Truly, the crew of STS-2. Courtesy NASA.

would drink had an awful lot of hydrogen in it, and once you got that into your system, it's the same way as when you drink a Coke real fast and it's still bubbly; you want to belch and get rid of that gas. That was the natural physiological reaction, but anytime you did that, of course, you would regurgitate water. It wasn't a nice thing, so we didn't drink any water. So we were dehydrated as well; tired and dehydrated when it was time to come back in.

In addition to the strained physical condition of the crew, other factors were complicating the return to Earth. The winds at Edwards Air Force Base were very high a couple of orbits before entry, when the crew was making the entry preparations, leading to discussion as to whether it would be necessary to divert to an alternate landing site.

Even under the best of circumstances, STS-2's reentry would have had a bit more pressure built into it—another of the test objectives for the second shuttle mission was to gain data about how the shuttle could maneuver during reentry and what it was capable of doing. The standard entry profile was to be abandoned, and about thirty different maneuvers were to be flown to see how the shuttle handled them. In order to accomplish this, the autopilot was turned off, making Engle the only commander to have flown the complete reentry and landing manually.

"Getting that data to verify and confirm the capabilities of the vehicle was something that we wanted very much to do and, quite honestly, not everyone at NASA thought it was all that important," Engle recalled.

There was an element in the engineering community that felt that we could always fly it with the variables and the unknowns just as they were from wind tunnel data. . . . Then there was the other school, which I will readily admit that I was one of, that felt you just don't know when you may have a payload you weren't able to deploy, so you have maybe the CG [center of gravity] not in the optimum place and you can't do anything about it, and just how much maneuvering will you be able to do with that vehicle in that condition? How much control authority is really out there on the elevons? And how much cross range do you really have if you need to come down on an orbit that is not the one that you really intended to come down on? So it was something that, like in anything, there was good, healthy discussions on, and ultimately the data showed that, yes, it was really worthwhile to get.

(In fact, after STS-2, maneuvers that Engle tested by flying manually were programmed into the shuttle's flight control software so that they could be performed by the autopilot if needed in the future.)

And so, with a variety of factors working against them, the crew members began the challenging reentry. "We had a vehicle with a fuel cell that had to be shut down, so we were down to less than optimum amount of electrical power available," Engle recalled.

The winds were coming up at Edwards. We hadn't had any sleep the night before, and we were dehydrated as could be. And just before we started to prepare for the entry, Dick decided he was not going to take any chances of getting motion sickness on the flight, because the entry was demanding. . . . He had replaced his scopolamine patch [for motion sickness] and put on a fresh one. The atmosphere was dry in the orbiter and we both were rubbing our eyes. We weren't aware that the stuff that's in a scopolamine patch dilates your eyes. So we got in our seats and got strapped in, got ready for entry, and I'd pitched around and was about ready for the first maneuver and said, "Okay, Dick, let me make sure we got the first one right," and I read off the conditions. I didn't hear anything back, and I looked over and Dick had the checklist and he was going back and forth and he said, "Joe Henry, I can't see a damn thing." So I thought, "This is going to be a pretty good, interesting entry. We got a fuel cell down. We got a broke bird. We got winds coming up at Edwards. We got no sleep. We're thirsty and we're dehydrated, and now my PLT*'s [pilot's] gone blind." Fortunately, Dick was able to read enough of the stuff, and I had memorized those maneuvers. That was part of the benefits of the delay of the launch was that it gave us more time to practice, and those maneuvers were intuitive to me at the time. They were just like they were bred into me.*

As the reentry progressed, Engle recalled, it felt like everything went into slow motion as he waited to execute one maneuver after another. And then, with the winds having cooperated enough to prevent *Columbia* from having to divert to another site, it was time for the former X-15 pilot's triumphant return to Edwards.

When we did get back over Edwards and lined up on the runway, as I mentioned before, I think one of the greatest feelings that I've had in the space program since I got here was rolling out on final and seeing the dry lake bed out there, because I'd spent so much time out there, and I dearly love Edwards and the people out there. In fact, I recall when Dick and I spent numerous weekends practicing landings at Edwards, I would go down to the flight line and talk with guys . . . and go up to the flight control tower and talk with the people up there, and we would laugh and joke with them. I remember the tower operator said, "Well, give me a call on final. I'll clear you." Of course, that was not a normal thing to do, because we were talking with the CapCom here at Houston throughout the flight. But I rolled out on final, and it was just kind of an instinctive thing. I called and I said, "Eddy Tower, it's Columbia *rolling out on high final. I'll call the gear on the flare." And he popped right back and just very professional voice, said, "Roger,* Columbia*, you're cleared number one. Call your gear." It caused some folks in Mission Control to ask, "Who was that? What was that other chatter on the channel?" because nobody else is supposed to be on. But to me it was really a neat thing, really a gratifying thing, and the guys in the tower, Edwards folks, just really loved it, to be part of it.*

According to Engle, the landing was very much like his experiences with *Enterprise* during the approach and landing tests.

From an airplane-handling-qualities standpoint, I was very, very pushed to find any difference between Enterprise *and the two orbital vehicles,* Columbia *and* Discovery, *other than the fact that* Enterprise *was much lighter weight and, therefore, performance-wise, you had to fly a steeper profile and the airspeed bled off quicker in the approach and landing. But as far as the response of the vehicle, the airplane was optimized to respond to what pilots tend to like in the way of vehicle response. . . . There were some things that would have been nice to have had different on the orbiter, . . . and that is the hand controller itself. It's not optimized for landing a vehicle. It really is a derivative of the Apollo rotational hand controller, which was designed for and optimized for operation in space, and since that's where the shuttle lives most of the time, it leans toward optimizing space operations, rendezvous and docking and those types of maneuvers.*

Despite the change in schedule caused by the shortening of the mission, more than two hundred thousand people showed up to watch the landing. After the vehicle touched down, Engle conducted a traditional pilot's inspection of his vehicle.

I think every pilot, out of just habit, gets out of his airplane and walks around it to give it a postflight check. It's really required when you're an operational pilot, and I think you're curious just to make sure that the bird's okay. And of course, after a reentry like that, you're very curious to know what it looks like. You figure it's got to look scorched after an entry like that, with all the heat and the fire that you saw during entry. Additionally, of course, we were interested at that time to see if the tiles were intact. . . . We lost a couple of tiles, as I recall, but they were not on the bottom surface. They had perfected the bonding on those tiles first, because they were the most critical, and they did a very good job on that. But we walked around, kicked the tires, did the regular pilot thing.

STS-3
Crew: Commander Jack Lousma, Pilot Gordon Fullerton
Orbiter: *Columbia*
Launched: 22 March 1982
Landed: 30 March 1982
Mission: Test of orbiter systems

The mission that flew after STS-2 could have been a very different one, had the shuttle been ready sooner or the sun been quieter, according to astronaut Fred Haise. The *Apollo 13* veteran had initially been named as commander of the mission, with Skylab II astronaut Jack Lousma as his pilot. The purpose of the mission would have been to revisit the Skylab space station, abandoned in orbit since its third and final crew departed in February 1974. At a minimum, the mission would have recovered a "time capsule" the final Skylab crew had left behind to study the effects of long-term exposure to the orbital environment. There were also discussions about using the shuttle mission to better prepare the aging Skylab for its eventual de-orbit. Unfortunately, delays with the shuttle pushed the mission backward, and an expansion of Earth's atmosphere caused by higher than predicted solar activity pushed the end of Skylab forward. "And what happened, obviously was there was a miscalculation, I guess, on the solar effect on our atmosphere, which was raised, causing more drag," Haise said. "So the Skylab [predicted time] for reentering was moving to the left in schedule, and our flight schedule was going to the right. So at a point, they crossed, and that mission went away."

Haise and Lousma trained for several months for the mission, and when the Skylab rendezvous became impossible, Haise decided to reevaluate his future plans. Given his own experience and the number of newer astronauts still waiting for a flight, he decided to leave NASA and take a management position with Grumman Aerospace Corporation. "I just felt it was the right time to start my next career. And so I left the program in '79 for that purpose."

Haise's departure meant changes for astronaut Gordon Fullerton, who recalled that at one point during shifting crew assignment discussions, he had actually been scheduled to fly with Haise on the second shuttle mission. "For a while I was going to fly with Fred. Then Fred decided he wasn't going to stick it out," Fullerton explained. "So then I ended with Vance [Brand] for a little while, and then finally with Jack Lousma, which was great. Jack's a great guy, [a] very capable guy and a great guy to work with, and so I couldn't have done better to have a partner to fly with."

Recalled astronaut Pinky Nelson of the pair:

Jack and Gordo were black and white. I mean, they were the yin and yang of the space program, basically. Jack is your basic great pilot, kind of "Let's go do this stuff," and Gordo is probably the second-most-detail-oriented person I've ever seen. Gordo at least knew he was that way and had some perspective on it, but there were things he could not let go. So he knew everything, basically. He knew all the details and really worked hard at making sure that everything was in place, while Jack looked after the big picture kind of stuff. They were a good team.

The shuttle was very much still a developmental vehicle when Lousma and Fullerton prepared for and flew STS-3. "When we flew STS-3, we had [a big book] called Program Notes, which were known flaws in the software," Fullerton explained. "There was one subsystem that, when it was turned on, the feedback on the displays said 'off,' because they'd gotten the polarity wrong . . . which they knew and they knew how to fix it, but we didn't fix it. We flew it that way, knowing that 'off' meant 'on' for this subsystem. The crew had to train and keep all this in mind, because to fix it means you'd have to revalidate the whole software load again, and there wasn't time to do that."

The big issue preventing the changes being made was time. The problems had been identified before the first launch, but continuing to work on them would have continued to delay the program. "They had to call a halt and live with some real things you wouldn't live with if you'd bought a new car. That's all part of the challenge and excitement and satisfaction that comes with being involved with something brand new."

While Fullerton was largely confident in the vehicle despite the problems, his only concern was the fact that the simulators were programmed assuming that the orbiter was working nominally. If a failure occurred in a system with a "program note," resolving it wouldn't necessarily look quite like it did in simulation.

It's really a complex vehicle. It really is. . . . If everything works like normal, it's all a piece of cake. It's when something breaks that you worry about, and the big challenge is to get to a point where you feel like you've got a handle on it. So was I ready to not show up on the launch date? No, not at all. Was I quaking in my boots? No. Was I intense about the whole thing? Yes, mostly because I am worried about my part of this. Especially for pilots, it's the launch phase [you worry about], because while it's short and concentrated, if anything goes wrong, the orbiter only takes care of the first failure. The second failure is pretty much left to the crew, generally, and so you worry about being ready to recognize a problem and do the right thing. You feel like the whole world's watching you when that failure occurs because of the manual action you've got to take to save the day. So it's that kind of pressure, pressure of performance, rather than fear or anything.

Lousma and Fullerton's flight built on the accomplishments of STS-2 in further testing the capabilities of the shuttle's Canadarm remote manipulator system. On STS-2, Truly had run the arm through a series of maneuvers without any added loads. On STS-3, Fullerton would take the next step by using the arm to grapple an object, lift it, move it, and then return it to place. Fullerton and Lousma paid particular attention to their physical condition during the mission, after the problems suffered by the STS-2 crew. "Jack and I worried about it a lot," Fullerton admitted. "One thing that we did do, that I don't think they did, is we had a g-suit, like they wear in the F-18, except that for entry you could pump up the g-suit and just keep it that way, and so that helped you keep your blood flow up near your head. . . . There was some controversy about whether you ought to pump them up or not, among individuals. We said, 'We're going to pump them up.'"

Another physical concern the crew worked to mitigate was orbital motion sickness. Fullerton noted that they looked into whether they could use NASA's T-38 astronaut jets to decrease problems with nausea in space.

We're not sure there's a direct correlation to flying airplanes and sickness. I know if you go up and do a lot of aerobatics day after day, you get to be much more tolerant of it. So Jack and I, we scheduled T-38s every chance we got in the last couple of weeks before we went down there, and I flew literally hundreds of aileron rolls. . . . If I did roll after roll after roll, I could make myself sick, and I did that, and I got to the point where it took hundreds of them to make me sick. But I did that figuring, I don't know if this helps, but I have the opportunity, I'll do it.

The results were pretty much the same on both of Fullerton's flights.

For the first day or so, I didn't ever throw up or anything. I never got disoriented, but I felt kind of fifty-fifty, you know. You're pretty happy to just float around and relax rather than keep charging. And into the second day, this is really fun and great, and you feel 100 percent. So whether the aileron rolls helped or not, I'm not sure, but it was relatively easy. Of course, everybody has their acclimation problems. That's pretty consistent through the population. It takes about twenty-four hours to get to feel normal, at varying levels of discomfort. Most everybody can hang in there and do their stuff, even though they don't feel good. A few are pretty well debilitated.

Pinky Nelson, who was a CapCom for STS-3, found it fascinating to observe a two-man team doing all the tasks necessary to fly the vehicle and also conduct mission objectives. "Flying the Space Shuttle with two people was a nontrivial job. It was a full-time job to keep that thing going with just two people and carry out some kind of a mission. I don't know how they did it, actually. I don't think I'd want to fly in the shuttle with just one other person."

Working as CapCom during that time was a challenge, Nelson explained, because the Tracking and Data Relay Satellite System didn't exist yet and the only opportunity to communicate with the astronauts was when the shuttle was over ground communications stations.

The time that you could communicate was very limited. You'd get a three-minute pass over Hawaii and a two-minute pass over Botswana or something, so you had to plan. Unlike now, when you can talk pretty much anytime, you had to plan very carefully and prioritize what you were going to say. The data came down in spurts, so the folks in the back rooms had to really plan for looking at their data and analyzing it and being able to make decisions based on spurts of data rather than continuous data. So it was kind of a different way to operate.

According to Nelson, there were some particular tricks to the art of being a good CapCom. One had to learn to speak succinctly and precisely and to stick to language that the astronauts were used to hearing from simulations. Another vital skill was listening to the tone of voice of the crew.

When you went AOS, *acquisition of signal, over a site, you would call up and say, "*Columbia, *Houston through Hawaii for two and a half," or something like that, and then you could just tell by the tone of their voice in the answer whether they were up to their ears or whether they were ready to listen. So there was a lot of judgment that had to be made, just in terms of, you always have a pile of stuff to get up. How much of this should I attempt to get up? What has to go up? Do I need to listen instead of talk? I found that to be just an interesting experience, a challenging job, and I really liked it. There were a few run-ins. I remember Neil Hutchinson, the flight director, was trying to get me to get a message up and I just wouldn't do it, because I knew that they just weren't ready to act on it, and it was important but wasn't critical or anything. And Neil was ready to kill me, and I just kind of sat there and just said, "No. They're busy. They don't need to do this now." So that was fun.*

Unlike STS-2, which was shortened by the fuel cell problem, Lousma and Fullerton's mission was actually lengthened by a day because of adverse weather at the landing site, Fullerton recalled.

"Wow!" We cheered. "Great!" because we really had a busy time with just two people. This was an engineering test flight, and we had a flight plan full of stuff . . . so there was always something that you were watching the clock on. . . . We did have sleep periods, which we would use for window gazing, . . . because you don't need as much sleep as they were scheduling. But when they said, "Wave off," I remembered getting in the recycle book, going through the pages, shutting down some of the computers, opening the doors again, and I got all the way down, all of the sudden, I turned the page, and there was nothing on it, and there was this realization, hey, this is free time, and it was terrific. We got out of the suits, and then we got something to eat and watched the world, and I wouldn't have had it any other way, if it had been my choice.

When the time for landing came the next day, the plan called for an early morning touchdown, meaning that the main part of the reentry would be at night. "We could see this glow from the ionization really bright out there," Fullerton said.

In fact, we had lost a couple of tiles on launch. We knew that because we'd looked out and had seen the holes in front of the windshield, and we looked at it with an arm camera. They said, "Not to worry. It's cool up on top there." We didn't know how many we'd lost from the bottom, but wasn't any use worrying about that. And then to see all this glow right there where the missing tiles were gave us pause to think about it. Again, there was no point in worrying about it, nothing you can do. [There was] the spectacular light show through entry. Then the sun came up, which washes all that out, as it's dying out anyway.

They were pushing at that time to go full-auto land, and so . . . we stayed in automatic all the way down through the pullout of the dive, and then [Lousma] only got the feel of the airplane the last couple of seconds before touchdown, which, in retrospect, everybody agreed was dumb, and now people fly from the time they go to subsonic at a minimum to get the feel of the airplane all the way down. He only got the last second, and then we landed a bit fast and . . . there was a kind of a wheelie that Jack did. Again, it pointed out another flaw or room for improvement in the software. The gains between the stick and the elevons that were good for flying up in the air were not good when the main wheels were on the ground, and he thought he had ballooned. He kind of planted it down but then came back on the stick, and the nose came up. So what? It didn't take off again, and we came down and rolled to a stop. A lot of people thought this was a terrible thing.

The landing at White Sands would leave a lasting mark on *Columbia*. According to astronaut Charlie Bolden, "I flew it several flights later, on my first flight, and when we got on orbit there was still gypsum coming out of everything. . . . It was just unreal what it had done." Astronaut Mike Lounge noted, "I'm told that many years later, picking up pieces from East Texas of *Columbia* [after the loss of the vehicle on the STS-107 mission], they were finding gypsum from White Sands."

STS-4
Crew: Commander T. K. Mattingly, Pilot Hank Hartsfield
Orbiter: *Columbia*
Launched: 27 June 1982
Landed: 4 July 1982
Mission: Test of orbiter systems

When the crews were chosen for each of the planned demonstration flights, Hank Hartsfield recalled, the astronauts were assigned originally not to a particular mission, but only to crews designated with a letter, A through F. "Ken [Mattingly] and I were in E crew. . . . No one knew exactly how this was going to work. All we knew was that Young and Crippen were A, and Engle and Truly were B. We knew John was first, and they were being backed up by Engle and Truly. But after that, we weren't quite sure what was happening. . . . Ken and I wondered, 'What are we going to fly?' . . . It was kind of a strange thing. Lousma and Fullerton were training. Eventually we figured out they were going to be [STS-3]."

Eager to figure out who was doing what, the astronauts in the remaining crews began paying close attention to what sort of training each was doing, looking for subtle clues. "We got this call . . . that we should go to St. Louis or wherever [the STS-3 crew members] were training; that Ken and I should go up there and start getting this training. It was kind of funny, because it scared them. Lousma made a panicked call back to Houston, said,

'What's going on? Are we being replaced?' Because nobody bothered to tell anybody what was going on."

As it turned out, Hartsfield explained, he and Mattingly were sent to train alongside the STS-3 crew so that they could be trained as a backup for the STS-2 crew while Lousma and Fullerton were preparing for their own mission. Hartsfield and Mattingly served as a backup for the third flight as well.

Then we flew [STS-]4. It was kind of a funny way the crews were labeled. . . . The D crew flew five, and we flew four, . . . the last of the two-person flights. It was a little bit confusing as to the way the crews were announced, . . . but it all sorted out, and I think sorted out fairly. Everybody got to fly, and nobody got kicked off a flight, you know. For some of us who had waited so many years—the seven of us that came from the MOL program, from the time we were picked for the space program till the time we flew was around sixteen years with the air force and a long time at Houston—it was a long wait. At that point, you didn't want to see anything get in your way. When the crew confusion started going on, "Well, I hope I'm not losing my place, I've waited too long." But everybody got to fly, so it was a good deal.

STS-4 commander Ken Mattingly recalled talking with Deke Slayton after the *Apollo 16* flight about what he wanted to do next. The two shared a relatively unique fascination with the shuttle program from a flight engineering perspective. "We both recognized that I enjoyed the engineering side of the flying, perhaps more than a lot of the guys. So the idea of trying to get in on an early flight test was what every pilot wants to do anyhow. The idea of being in a group that was going to be downsized and have an opportunity to participate in the first flights and maybe even compete for the first flight, that was all the motivation anybody could ever want."

While Mattingly was not surprised when the first flight went to his *Apollo 16* commander and the corps' senior flight-status astronaut, John Young, he was disappointed that he had to wait until the last development mission to get to fly. The rationale, he explained, was that his expertise was needed in different ways—because his flight opportunity was delayed, Mattingly was available to back up the second and third crews should something go wrong, and then, on the fourth flight, he could complete any tasks that hadn't been accomplished by the first three crews. "That was the logic. It was kind of fun to be part of those missions, but . . . Hank and I were kind of hoping we could [fly] earlier. But it really did turn out to have a lot of benefits for us, because we did pick up a lot of experience we would not have had and were able to do some other activities that [we] wouldn't have had time to go do if we'd been scrambling just to get up and down."

Although Mattingly and the rookie astronaut Hartsfield had never flown together before, their assignment to STS-4 was a reunion of sorts: Hartsfield had served as CapCom during Mattingly's Apollo flight to the moon.

In [Apollo] 16, *Hank and I had developed a better-than-average rapport, I think, because in lunar orbit, Hank ran the show and all the flight plan from the ground. I told him, "You only get to go to the moon once, so I don't want to miss a minute of looking out the window. So you run the spacecraft, and I'll look and tell you about it." And he really did a magnificent job on that, and as a result, we got a lot of stuff done that we wouldn't have otherwise. So on* STS-*4, it was kind of fun to go back to working together that way, and we were still trying to see how much we could cram into this thing.*

Hartsfield added that not only did the two astronauts get along well, they had a unique commonality. "We both went to Auburn. I think it's the only time the entire crew went to the same university in the spaceflight business. We used to take a lot of ribbing from the University of Alabama folks, saying the only reason they put two Auburn guys on one flight, that way we'd only mess up one. So we had to take a lot of ribbing, but it was a lot of fun."

STS-4 CapCom Pinky Nelson described having an interesting working relationship with the STS-4 commander. "T. K. Mattingly is probably the most technically capable person who has ever been an astronaut, just in terms of his capacity to stuff things between his ears," Nelson voiced.

He knew absolutely everything, and had to know everything, and was fanatical about tracking everything, and drove me nuts, because I don't work that way. I tend to work in a way where you take in a lot of information, but you have a filter. You say, okay, this is important, this might be important, this is probably not important, and you prioritize things, where T.K. works that everything is on the top line. He's able to work that way just because of his incredible capacity, and I wasn't, so he and I had kind of an odd relationship. If I didn't see the point of having to do something, I wouldn't do it, basically.

Despite their differences, Mattingly and Nelson had enough in common, and enough willingness to coexist where they were different, that they actually had a decent working relationship.

T. K. and I really got along. We were able to communicate fairly well because we had the same kind of style of no extra words kind of communication. But, boy, he was a hard taskmaster, I thought. He just didn't see the forest, but he saw every tree, and he expected everybody else to do that, and they just couldn't. He really wore some people down. That kind of thing doesn't bother me so much. I was able to just kind of ignore it and say, "Okay, I'm going to catch some flak for this, but I deserve it. I don't care. I'm not going to do it anyway." . . . Some newspaper article about the mission called me the "laconic and taciturn CapCom." That was great.

When launch day came, Hartsfield was excited that, after sixteen years in the air force and NASA astronaut corps, he was finally going to fly.

To me, it was kind of an emotional thing. I remember when we were going out to the pad in the van, and just before we got up to the pad to get out and go to get in the bird, it just sort of hit me, and I said, . . . "Ken, I can't believe it. I think we'll really get to do this." It hit me emotionally, because tears started welling up in my eyes. You know, I had to wipe my eyes. It just, to me, was an emotional thought, after all that time, I was finally going to get to fly, it appeared. And I did.

Mattingly's perspective on the ascent was different from that of many of his peers, as he was one of only a few astronauts who could compare launch on the large Saturn V rocket and the Space Shuttle. Many shuttle astronauts have commented on the power and vibration of the solid rockets during the first two minutes of flight; Mattingly, on the other hand, noted how relatively smooth the launch was. "Compared to the Saturn, the shuttle is like electric propulsion; it doesn't make any noise, it doesn't shake and rattle, it just goes. It's just nothing like the Saturn, or, as I understand, the Gemini or the Titan."

Hartsfield recalled that the launch was not without complications, due to a hailstorm the night before.

About nine o'clock that night, after the storm passed over and it quit raining, we went out to the pad with a lot of other people to look at the orbiter. And the black tiles all had little white specks all over where they'd been pelted with the hail. So Ken and I went back to the crew quarters thinking, "Shit, we ain't go-

ing nowhere tomorrow." So we were real down. And we went to bed and couldn't believe it the next morning when these guys were hammering [on] the door, saying, "Come on, get up, guys, we're going." We said, "What, we're going?" "Yeah, they cleared it." They flew some guy from Houston in a T-38 down there, one of these tile experts, and he went out and walked around and decided that it was okay to go, that the tiles weren't damaged that badly.

The next day it was discovered that there was a side effect of the damage that the expert had failed to anticipate. During launch, controllers noticed that the vehicle was not getting the performance it should for the amount of fuel being used. It turned out the hail had damaged the waterproof coating of the tiles, and they'd absorbed water during the storm. "They calculated later that we'd carried about two thousand pounds of water with us that had soaked into the tiles," Hartsfield explained. "So all this flight planning that we'd worked on for so long and had down pat went out . . . the window. We spent seventy-two hours with the belly to the sun trying to bake out the tiles. Because they were concerned that if you had water in those tiles and then got entry heating, they didn't know whether you might get some steam or something generated and blow the tiles out or crack them or something."

With the excitement and drama of launch done, the arrival in orbit was a moment of wonder even for veteran astronaut Mattingly.

The most magical thing was, after working on this device for ten years, you got on orbit and . . . we opened the payload bay doors for the first time towards the Earth. So all of a sudden, it was like you pulled the shades back on a bay window, and the Earth appeared. We got on orbit, and this thing worked. And I just couldn't get over the fact that . . . people that I knew, that were friends, had built and conceived this whole thing, and it works. It's just magic. It does all of these things that we dreamed of, but the visuals are better than the simulator now. So we just had a wonderful time of it.

Flying around the Earth is just so spectacular. I don't care how long you're up there, I can't imagine anyone ever getting tired of it. It's just beautiful, and the orbiter with these big windows, it is just wonderful. Hank would say, "You know, we probably ought to get some sleep here." I'd say, "Yeah, yeah, yeah. You're right. We've got another day's work tomorrow." . . . So all the kids are in bed, and now you can look out the window. I told the ground I went to sleep so they won't bother me, and I'd sit there, having a wonderful time.

When he finally got tired enough to stop looking out the window and try to get some sleep, Mattingly decided to see what it would be like to sleep freely, instead of hanging from a wall in his sleeping bag. Since Hartsfield was sleeping on the mid-deck, Mattingly had the flight deck to himself, and he decided to try just lying down on the floor.

I worked at getting all steady and not moving and stopped right behind the two seats that had a little space over the hatches that come up from the mid-deck and in between the aft control panel and the back of the ejection seats, which there's a lot more room today since they took the ejection seats out. So it was a place probably two feet wide, maybe two and a half. I got all stable in there. "Ah, this is nice. Go to sleep." Well, the next thing I know, there's something on my nose, and it's a window, and god-dang, I was sure I had gotten stable. So I went back and set up again, not moving, did it again, ended up with my nose in the window, in the overhead window. That bothered me. I finally put a Velcro strap over me just to keep me from floating up. I just thought that was really curious. So the next morning I was telling Hank about it, and he said, "Well, I didn't have any trouble. I just was floating in the middle of the mid-deck." Hmm.

Hartsfield pointed out that during the sleep shift the orbiter had been carrying out passive thermal control maneuvers. In order to better understand the thermal characteristics of the vehicle, flight controllers would change its orientation to expose different parts of it to sunlight for periods of time, taking advantage of instrumentation that wouldn't be used after this last development flight.

"[Hartsfield] says, 'You know, I was almost on the center of rotation, and you were up here. This is centrifugal force,'" Mattingly recalled.

I said, "Oh, come on, Hank. What was it, . . . five revolutions an hour, or some gosh-awful thing? . . . That can't be." He said, "Well, we've got another one scheduled for tonight. Let's try it both ways." We tried it, and sure enough, every time. If this thing was rotating at this really slow rate, there's no other force; these little forces become important. And after we stopped, he says, "Try it again." I did, and sure enough, no problem. So this is kind of added to some of the little micro physics things that you see in space that are so interesting.

In addition to the standard tests of the orbiter systems, STS-4 was the first shuttle flight to include a classified military experiment. In order to preserve secrecy, the experiment had its own classified checklist with coded section names that could be discussed over the unsecure communications channel. The experiment itself was kept in a padlocked locker. While the shuttle was in orbit, the crew members could leave it unlocked, but once the experiment was completed, they had to stow it and lock it back up to keep it secure after landing. Hartsfield recalled that about thirty minutes after they finished the experiment and locked it away, they got a call from the military flight control for the experiment.

The CapCom came on, the military guy, and says he wanted me to do Tab November. Ken said, "What's Tab November?" I said, "I ain't got the foggiest idea. I'm going to have to get the checklist out to see." So I got the padlock off and got the drawer and dug down and got the checklist out and went to Tab November, and it says, "Put everything away and secure it." Ken and I really laughed about it. It was just aggravating to have to undo all that, because that locker, the stuff we had just barely fit in there, so it was really a stowage issue here. If there's one thing you learn in zero g—things are always neatly packed [before flight] and you get it up there, and once you pull it out, it doesn't always go back in, because it expands or does something in zero g and it doesn't fit very well.

Like the first three flights, STS-4 involved experiments to test the capabilities and tolerances of the orbiter. One particular experiment, testing the thermal tolerances of the payload bay doors, resulted in a tense moment for the crew. In the experiment, the orbiter was kept in a position to expose the doors to the sun for three days.

"After seventy-two hours to the sun, it came time to cycle the payload bay doors," Hartsfield said.

We brought the port door down. We're looking out, and the door comes down. And all of a sudden, the left collar of that thing hits the bulkhead and the door just warps. And by the time I got it stopped, it's already done. And we tried to call the ground and say we've got a problem here, but about that time, we went LOS. [On the ground] they're panicked. They see what's happened, that the door is hung up on the bulkhead or something, and the door's warped. And you know, [JSC director Chris] Kraft was not a flight director, but he sat in the back watching everything. "Tell 'em to open the door. Tell 'em to reopen the door! Tell 'em to open the goddamn door!" They tell me he was just getting furious. . . . And they couldn't get it to us. So Ken and I were saying, "Holy shit, that door is really bent."

Hartsfield and Mattingly wondered if the door was broken, which would create a catastrophic situation. During entry and landing, the structural integrity of the orbiter required the doors to be closed and all but one set of the latches to be latched. "So we were wondering if we'd broken something and weren't going to able to latch the doors back up to come home. Well, as soon as we got AOS again, they told us to open the door, and I started driving it, and it all of a sudden—boing—the door vibrated and it went back to its normal shape. And we went, 'Whew.' . . . Thank God it's a composite material, so it does have some kind of resiliency to keep its shape once you take the load off of it."

While earlier flights had landed in the dry lake bed at Edwards or at White Sands, STS-4 would be another stepping stone for the shuttle program by being the first to return from space on an aircraft runway, and Mattingly was hoping to avoid a repeat of Haise's infamous bounce landing of *Enterprise*. "Our job, like Freddo's, was to plan to make the first concrete runway landing. You know, as much as we trained for that thing, I just had this image of doing Freddo's trick all over again. It was, you know, bad karma or something. Oh, that bothered me. I could think of nothing else."

Mattingly had promised Hartsfield that he would let the pilot fly a part of the entry so that he could say he had actually flown the orbiter, normally the sole privilege of the commander. "When we did come in and got out at Edwards and came around," Mattingly said,

we got on the heading alignment circle and I was tracking it, and I turned to look at Hank, and I was about to say, "Well, okay, here, you take it for a bit now," . . . and all of a sudden my gyros tumbled and I just had one of the worse cases of vertigo I've ever had. . . . It was just really overwhelming. I went back and started focusing on the eight ball and looking at the displays, and Hank says, "Are you going to let me fly?" And I said, "No, no. I can't talk about it now." And we came around and did our thing, and I was still having this vestibular sensation that was unusual, but once we got on the glide slope it seemed like . . . normal.

As they neared the runway, Mattingly said because he felt a little off, they were a little slow going through procedures leading up to the flair point, and they ended up flying under the standard approach.

I knew we were under the standard final approach glide slope, but now I wanted to get down and try to make a good landing with it. . . . So he's calling off airspeeds and altitude, and I'm just staring at the horizon and I'm hawking it, and I have no idea what it's going to feel like to land. When I would shoot touch-and-gos in the [KC-135 aircraft], there was never any doubt when we landed. You could always tell. So I was expecting bang, crash, squeak, something. Then nothing and nothing. Then finally Hank says, "You'd better put the nose down." "Oh," I said, "all right." So I put it down, and I was sure we were still in the air. I thought, "Oh, God, he's right. We can't be very far off the ground." Sure enough, we were on the ground and neither one of us knew it. I've never been able to do that again in any airplane. Never did it before. According to pictures, it looks like we must have landed at maybe 350 feet down the runway, and we didn't mean to.

During planning for the flight, it was made very clear that even after the orbiter was safely stopped on the ground, there would still be one important mission objective for the astronauts—a proper patriotic performance. "We knew that they had hyped up the STS-4 mission so that they wanted to make sure that we landed on the Fourth of July," Mattingly noted.

It was no uncertain terms that we were going to land on the Fourth of July, no matter what day we took off. Even if it was the fifth, we were going to land on the Fourth. That meant, if you didn't do any of your test mission, that's okay, as long as you just land on the Fourth, because the president is going to be there. . . . The administrator met us for lunch the day before flight, and as he walked out, he said, "Oh, by the way . . . you know, with the president going to be there and all, you might give a couple of minutes' thought on something that'd be appropriate to say, like 'A small step for man,' or something like that," and he left. Hank and I looked at each other and he says, "He wants us to come up with this?" And we had a good time. We never came up with something we could say, but we came up with a whole lot of humor that we didn't dare say.

After landing, the crew members prepared for their moment in the spotlight. Anticipating that the president might want to come aboard the shuttle, Mattingly recalled, they put up a handwritten sign that read, "Welcome to

Columbia. Thirty minutes ago, this was in space." The astronauts took off their helmets and started to get out of their seats, a task Mattingly found surprisingly difficult after having adjusted to a microgravity environment.

I said, "I am not going to have somebody come up here and pull me out of this chair. . . . I don't know what it is, but I'm going to give every ounce of strength I've got and get up under my own." So I . . . pushed, and I hit my head on the overhead so hard, the blood was coming out. Goddamn. It was terrible. Oh, did I have a headache. And Hank said something like, "That's very graceful." So now I really did have something to worry about. . . . Hank's got some of the funniest stories he could tell about this stuff. So we got ourselves down there, and we're walking around, and Hank said, "Well, let's see, if you do it like you did getting out of your chair, you'll go down the stairs and you're going to fall down, so you need to have something to say." He says, "Why don't you just look up at the president and say, 'Mr. President, those are beautiful shoes.' Think you can get that right?" He was merciless.

7. Open for Business

The demonstration missions were complete. The shuttle was considered operational. America's new Space Transportation System was open for business, and one of its primary purposes would be the deployment of satellites.

STS-5
Crew: Commander Vance Brand, Pilot Bob Overmyer, Mission Specialists Joe Allen and Bill Lenoir
Orbiter: *Columbia*
Launched: 11 November 1982
Landed: 16 November 1982
Mission: Launch of two communications satellites

Satellite deployment was precisely the mission of the first operational mission, STS-5. It was the first shuttle flight to deploy satellites into geosynchronous orbit, which is ideal for communications satellites because the satellite remains at roughly the same longitude as it orbits Earth. To achieve a geosynchronous orbit, a satellite must reach an altitude around one hundred times higher than that at which the shuttle orbits. For the shuttle to launch such a satellite, the satellite was carried into low Earth orbit in the shuttle's payload pay and then boosted into the higher orbit using a booster rocket.

STS-5 was a mission full of firsts. In addition to the satellite deployments, STS-5's four-man crew doubled the crew from the previous missions' commander and pilot team to include the first two mission specialists to fly, Joe Allen and Bill Lenoir. Because of the deployments, STS-5 was also NASA's first "commercial mission," with the focus being on performing a service for paying customers.

Joe Allen had been selected as part of NASA's second group of scientist-astronauts in 1967, and the move to larger crews had long been awaited by him and his classmates. "The first assignment of mission specialists we knew was going to be [STS-]5, because the system was going to be declared operational after the first four test flights if nothing untoward happened," Allen recalled.

We also knew that the next in line to be assigned were the scientists-astronauts. Those who had arrived [in the first of the two scientist-astronaut selections] had already flown, Jack Schmitt being the first to fly on the last Apollo, and then Joe Kerwin, Ed Gibson, and Owen K. Garriott had flown on the Skylab. So there were just now, I think, nine of us who would be considered. I guess I never thought much about it, but I almost assumed that maybe they went alphabetically, because they put myself and Bill Lenoir aboard that first operational flight. And I was thrilled, absolutely thrilled.

After having been assigned to the crew for just a few weeks, Allen decided he understood why he had been picked to be a part of that particular team, and he commented on it to George Abbey, who made crew assignments. "I was the impedance matching device between the two marine pilots and the MIT engineer. 'Impedance matching' is an engineering term for getting very unlike electrical circuits to communicate, one with the other."

Allen recalled that while he made the comment somewhat in jest, Abbey did not find it as amusing as he did. "I suspect that there were elements of truth in this, because I was very good at getting different groups of people to understand each other. With no scintilla of modesty at all, I would say that's probably my strongest suit, understanding the way different individuals think about things and then enabling communication between them, in spite of their differences. I assert I was—I hope it's not overblown—very successful in getting scientists to understand what flight-crew members needed, and getting flight-crew members to understand what scientists needed, even though neither group spoke the other's language."

Both groups had the same motivation—a successful mission—but each approached the task in its own way. "You wanted the ultimate result to be a successful mission, and successful in later Space Shuttle flights and in the last Apollo flights meant scientifically rich in what was achieved. So I was the impedance matching device between Bill Lenoir, an extremely smart, very well disciplined, very tightly wound individual, and Vance Brand and Bob Overmyer, whose backgrounds were military, with a military way of thinking about things, and they had a much higher tolerance for people being not quite so intense."

The addition of mission specialists, and the corresponding increase in crew size, resulted in a modification to the orbiter for this flight. Prior to the mission, the commander's and pilot's ejection seats were pinned to prevent them from being able to be used. In *Space Shuttle "Columbia": Her Missions and Crews*, Ben Evans wrote that the crew cabin was not large enough to have ejection seats for every member of the crew, only the commander and pilot. The thought with STS-5 was that if all of the crew wouldn't be able to eject in the event of a problem during launch, none of the crew should be able to do so. Allen recalled that he and Lenoir were opposed to the pinning, arguing that it would be better for two astronauts to survive than for none to, but Brand disagreed. "He said, 'That's not a choice.' He later said to me, 'Joe, this is not a selfless decision on my part; indeed, . . . it's selfish, because I could not live the rest of my life knowing that I survived and you didn't. I couldn't do it. I don't think Bob could either. . . . I have some historical evidence as to that being a true statement. I don't just surmise it.'"

Allen explained that Brand had been a test pilot in England for a while, and some of the English bombers enabled the pilots to get out, but not the gunners, so there was a small body of data from psychological studies done on individuals who had escaped but, in escaping, had left their shipmates to a certain death. "They had been definitely tormented, in terms of what the data showed, for the rest of their lives. So it was not a good solution. Vance was aware of that data, and he didn't want to be another bit of statistic in that database—a decision I thought was gracious. He said it was selfish; I didn't think it was selfish at all."

When launch day finally arrived, STS-5 added to its list of firsts by becoming the first shuttle mission to launch right on time. "We didn't have one hiccup, not one delay, no nothing, and we went . . . on time," recalled Allen. "We were told the night before that a Russian spaceship had actually changed the timing of its orbit somewhat and would come right over the Cape at exactly that time. Because we did launch on time, I suspect there are some photographs that could be found in the archives of the Soviets of us coming off the ground. I have no idea, but clearly they intended to at least watch us do it with their own eyes."

On the day before launch, however, Leonid Brezhnev, the premier of Russia, died, and newspapers around the world were filled with news of his death, not of the Americans' space launch. "There was maybe a little blurb, a little photo of us in the lower corner of some newspapers, but we were pretty much second-page news, with the exception of my hometown, Crawfordsville, Indiana, where I was the front page," Allen said.

That hometown front page was a major step forward for Allen. Years earlier, in August 1967, when he was selected as an astronaut, this same hometown newspaper ran a front-page account of the calf-judging contest at the county fair, with only a small article on the back page about the local native selected to be a NASA astronaut. "My mother knew the newspaper editor very well, and she called him, quite upset," Allen recalled. "She said to him, 'Harold, you know my son Joe was selected as an astronaut. I think that's very important news, and there's practically nothing.' And he said to her, 'Harriet, you know perfectly well we're a small town; we're a very small newspaper. If you want a story about your son Joe in the paper, you're going to have to write it yourself.'" By the time he actually flew into space, however, Allen merited front-page news without his mother having to write it.

The first priority of STS-5 was to successfully deploy the first hardware put into orbit by the Space Shuttle. Two commercial communications satellites were deployed, one for Canada and one for Satellite Business Systems. Allen said the crew worked closely with the satellite developers to understand how the satellites were put together and what was needed for successful deployment. "The concept of the satellite in itself is simple," Allen explained.

They are meant to be deployed spinning, and the way you do it is just put the satellite on a table that will spin, like you put a record onto a record player. . . . It's mounted on the table prior to launch, then you go to orbit. . . . Once there, you cause the table to spin and you point the shuttle in exactly the right direction, and then at precisely the right part of the orbit, you just release hold-down arms that are holding the satellite to the tabletop. When you release the arms, springs on which the satellite sits expand and just give it a very gentle push out, spinning very beautifully.

The second-highest-priority planned objective of STS-5 was the Space Shuttle's first extravehicular activity, by Allen and Bill Lenoir. "Although we had other bits and pieces of experiments to do," Allen recalled, "on the day before the spacewalk was to take place, we commented to ourselves that we really had just two important things left to do. One was the spacewalk and then the second was the reentry and a safe landing. I made the observation, 'Vance, out of these two, if we have to make a choice, let's choose the safe landing.' And we all laughed about that."

But what started as a joke quickly turned into reality when Allen encountered a problem with his suit. "Just as the spacewalk was to begin my spacesuit failed," Allen said.

It was an electrical failure in the spacesuit. When one is in a spacesuit, and you power it up, you hear a very high pitched hum there someplace in your ear, just a high-frequency hum. When I powered mine on, the hum started, but it didn't sound like it was healthy. It sounded, indeed, more like an angry mosquito. It just changed its pitch. I'd never heard it before. Then I proceeded to make various electrical checks of the suit systems, and none passed the check. So we tried all sorts of things, powering it on and off, and it was just not going to work, a very bitter disappointment to me, without any question. It was equally bitter to Bill, and the question now was, could he even do just a little part of a spacewalk—a short solo, if you will. . . . But Mission Control said, "No buddy system, no spacewalk." Bill was really upset, obviously.

After the crew's return to Earth, engineers analyzing the failure determined that the suit had a serious electrical problem but one that could be easily repaired for future spacewalks. "The bad news was the spacesuit failed; the good news was we were not outside the ship when it failed," Allen noted. "It would have been considerably more traumatic had it failed outside. It would not have been fatal to me, but it would have gotten my attention, for sure. I would have to scramble to get back in, button myself up, and get out of the suit before other parts of it started to fail."

Brand recalled how, as commander, he made the call that cost his mission its place in history as the first shuttle with spacewalks: "I guess I was the bad guy. As much as I hated to, I recommended to the ground that we cancel the EVAs, because we had a unit in each spacesuit fail in the same way. . . . We could have taken a chance and . . . could have done it, the EVA, but we didn't. I'm not sure Bill Lenoir was ever very happy about that, because he and Joe, of course, wanted to go out and have that first EVA."

For the first four-person crew and the largest yet to inhabit the orbiter, living aboard the spacecraft for the duration of the five-day mission created interesting challenges when it came to sharing the small amount of habitable space. For example, there were no sleeping compartments, Brand said, so they had to be creative in each finding a place of his own for rest. "Since we didn't have a sleep station, I just would take a string and tie it to my belt, and tied the other end to the wall. Tethered by the string, I'd put on a jacket and just fall asleep. It was great in weightlessness, and I slept well."

Bob Overmyer used the only sleeping bag and slept in the lower deck of the Space Shuttle, while Bill Lenoir slept in a corner on the upper deck. "I don't know how he did it, but he didn't float out of the corner, and he would sleep that way," Brand said. "Joe Allen was the funniest. He would just free-float through the spacecraft the whole night, or what we called night. He moved gently, and the air currents would move him around. He might gently bump into a switch or something, but they were covered and he was hitting them so gently that there wasn't really any concern about moving the switches. That was really humorous, seeing that."

As with the demonstration flights, mission objectives included checking out the orbiter systems and determining their capabilities and tolerances. One such task, Brand explained, was to test the orbiter's thermal tolerances by positioning the vehicle so that one side was fully in the sun and the other side was in shadow. "Sometimes after spending many tens of hours in one attitude, the shady side of the ship would get real cold," Brand recalled. "Dew would form on the inside of the ship on the cool side. We had a treadmill in the mid-deck, and when people ran on the treadmill, it shook that dew loose and it would sort of rain inside the spacecraft."

Allen, who would later publish a book featuring on-orbit photography, explained that NASA regulations almost prohibited him from taking a historic photo of the first time four Americans were together in space. In preparing for the mission, he discovered that the delayed shutter release button on his camera had been removed.

I went back to the photo people and I said, "This is a defective camera. It's been plugged. I want a real Nikon." Well, it had been modified, and the modification had cost tens of thousands of dollars in order to make it more astronaut-proof, such that astronauts didn't, by mistake, put the camera on delayed timing and thus mess up a picture. A couple of modifications had been done, none of which was necessary, all of which were costly, and I just was very upset with my colleagues. I said, "Cameras are amongst the best-tested of complicated tools that we have humans have. They've been tested in wars. They've been tested in violent storms. They've been tested at the bottoms of the oceans. They've been tested everywhere. Why are you making them better?" Well, it was an argument I did not win.

Not to be deterred, Allen was determined to find another solution to the issue.

Perhaps in spite of the rules and because I've been a little bit of a troublemaker, but not serious trouble, I went to a camera store and was able to find a very old-fashioned shutter release mechanism. . . . I took this device aboard the spaceship without anybody really knowing it, and it came secretly off the spaceship on my person. This was against NASA regulations, and I will readily admit to it. But in the flight photos that came back, there were numbers of photos of us, the four crewmen. [One was] good enough to appear in Time *magazine the next week. But this was with a camera that had no delayed shutter release! Not one NASA person said a word to me about it, but you knew that the people in the photo shops wondered how in the world those photographs were taken. A very nice man ran NASA Photo for many years; when he retired, I gave him that secret shutter release device—a flown object. . . . Because I knew he knew how I'd made that photo. He just had to know how I'd done it.*

At last, it was time to return to Earth. Brand, who had previously experienced reentry in an Apollo capsule at the end of his Apollo-Soyuz mission, said his return to Earth on the shuttle was a very different experience.

There were very large windows, and you weren't looking backwards at a donut of fire. You were able to see the fire all around you; you could look out the front. First the sky was black, because you were on the dark side of the Earth, but as this ion sheet began to heat up, you saw a rust color outside, then that rust color turned a little yellowish. Eventually, around Mach 20, you could see white beams or shockwaves coming off the nose. If you had a mirror—and I did on one of my flights—you could look up through the top window, which was a little behind the crew's station, and see a pattern and the fire going over the top of the vehicle, vortices and that sort of thing.

The crew landed at Edwards Air Force Base at 6:33 on the morning of 16 November 1982. "We had an intentional max braking test and completely ruined the brakes," Brand said. "I had to stomp on them as hard as I could. . . . Even though it was billed as the first commercial flight, I think we had roughly fifty flight-test objectives. That braking test was just one of them. We ruined the brakes, completely ruined them, but it was a test to see how well they would hold together if you did that."

STS-6
Crew: Commander Paul Weitz, Pilot Bo Bobko, Mission Specialists Story Musgrave and Don Peterson
Orbiter: *Challenger*
Launched: 4 April 1983
Landed: 9 April 1983
Mission: Deployment of TDRS-1; first flight of *Challenger*; first shuttle spacewalk

Paul Weitz, Bo Bobko, Story Musgrave, and Don Peterson were not, at first, officially the crew of the sixth Space Shuttle mission. Like the members of the other early crews, they were given a letter designation, F, when they participated in the development of the Orbiter Flight-Test program. The crew members nicknamed themselves "F Troop," a reference to a television comedy series featuring an Old West army troop. The crew even took an F Troop–themed crew photo. "We had on the little flight T-shirts and the flight pants," STS-6 mission specialist Don Peterson recalled,

but we went out and bought cowboys hats. I had a sword that had once belonged to some lieutenant in Napoleon's army. We got a Winchester rifle, the lever-action rifle, and a bugle and a cavalry flag, and we posed for this picture. Weitz, of course, is the commander, and he's sitting there very stern looking, with the sword sticking in the floor. I had the rifle, and I think Story had the bugle. Anyway, we had that picture made, and we were passing them out, and NASA asked us not to do that. They thought that was not dignified, but I thought it was hilarious. I still have a bunch of them.

While the crew members embraced the F Troop nickname, there was another nickname used mainly behind their backs. "I didn't hear 'the Geritol bunch' until, I guess, after the flight was over," recalled STS-6 commander Weitz, who had been selected as a member of NASA's fifth astronaut class seventeen years earlier. "Maybe that was something that everybody said about us when we weren't around. We were on orbit and somebody was talking about 'how old you guys are.' We had taken a bunch of pictures, and I couldn't resist, I said, 'You know, we're not going to show the pictures to anybody under thirty-five when we come back. So some of you guys, some of you wiseasses, won't see them.'"

STS-6 had three major mission objectives. It was the maiden launch of the second operational orbiter, OV-099, better known as *Challenger*, and so the crew would be making sure the new vehicle operated properly. The flight was to make the first deployment of a Tracking and Data Relay Satellite, part of a new space-based communications system. And, after the problems with the spacesuits on STS-5, this flight would now be making the first shuttle-based spacewalk.

Peterson and Musgrave were responsible for the deployment of the TDRS-1 satellite on the first day of the mission. "Story's the kind of guy that he wants to throw the switches," Peterson recalled, "so what I did was took the checklist. Story was not real good about following the checklist, and so you had to kind of say, 'Wait, Story. Let's go step-by-step here and make sure we get this right.'"

A few days before launch, while the crew was quarantined in the crew quarters at Cape Canaveral, Peterson and Musgrave were informed that changes had been made to the software they would use to deploy the satellite. "These two guys showed up and . . . said they were from Boeing. They had badges. . . . Story and I literally copied a bunch of stuff down with pen and ink and used that on orbit," Peterson said. "And that's really scary, because, you know, you're taking these guys' words. You've never seen some of this stuff in the simulator. It's, like, suppose what they're telling us is not right, and we do something and we mess up the payload. Then they ain't never going to find those two guys again. They'll be gone, and it'll be, like, 'Why the hell didn't you guys do it the way you were trained to do it?'"

The deployment went as planned, but after deployment, one of the two solid rockets in the booster that was to transport the satellite from the shuttle's orbit to its intended geosynchronous orbit failed. However, NASA was able to nudge the satellite into its proper orbit using extra fuel in the attitude control system that allowed the orientation of the satellite to be changed.

Bobko explained, "Luckily, they had planned to use the satellite for commercial purposes as well as NASA, and it was decided not to do that, but they didn't make that decision in time to take off some of the extra fuel that

19. The deployment of the first Tracking and Data Relay Satellite, on STS-6. Courtesy NASA.

was required for using the satellite commercially. So that fuel was available, and, luckily, that was what saved the satellite."

Musgrave and Peterson also were assigned to make the first shuttle-based spacewalks. "It's kind of funny," Peterson said. "George Abbey, I think, had some people already picked out that he wanted to have the honor of doing the first spacewalk, and when that canceled, he said, 'Well, we'll have to slip now. It'll take months to get another crew ready.' Jim Abrahamson, who's an old friend of mine, was the [associate] administrator. He called me on the phone and said, 'Can you and Story do the spacewalk?' And I said, 'Yeah.' So he said, 'Okay. We're going to do it on the next flight.'"

The late addition of the spacewalk didn't give the astronauts much time to train. Peterson had very little experience in the extravehicular activity spacesuit, but Musgrave had represented the Astronaut Office in the suit's development so he knew everything about the suit there was to know, Peterson said. "Story had spent, like, four hundred hours in the suit in the water tank, so he didn't really have to be trained," recalled Peterson. "Now, my training was pretty rushed, pretty hurried. I think I was in the water, I don't know, fifteen, twenty times, but that's really not enough to really know everything you need to know. But all we were doing was testing the suit and testing the airlock, so we weren't really doing anything that was critical to the survival of the vehicle. We were just testing equipment, and the deal was, if something went wrong, you'd just stop and come back inside. So the fact that I wasn't highly skilled in the suit really didn't matter that much."

Back during the Gemini program, NASA had determined that the best way to prepare for spacewalks was to make suited dives in a water tank. The spacesuit is weighted to make it "neutrally buoyant," such that it doesn't sink to the bottom or float to the surface; it just hangs there. The simulation of weightlessness isn't perfect—inside the suit, the astronaut's body is not floating; it's supported by the suit. So if the astronaut turns upside down in the water tank, the weight is on his or her head and shoulders.

In orbit, because of the difference between the pressure inside the shuttle (approximately 15 pounds per inch) and the pressure inside the suit (4.3 pounds per inch), the spacewalkers needed to slowly adjust to the pressure in the suit to prevent them from getting what's commonly called "the bends," a condition caused by too much nitrogen in the blood. To purge their bodies of nitrogen, the astronauts breathed pure oxygen for approximately three and half hours. "While we were breathing oxygen for three and a half hours, you can't really do anything," said Peterson. "Story and I slept. I slept about two and a half hours, probably the best sleep I had on orbit, because you've got fresh oxygen coming in over your head, and it kind of makes a nice whishing sound, and there's no other noise. . . . People asked, 'How in the world can you sleep just before you're getting ready to go?' I said, 'Well, you know, you get tired enough, you can sleep almost anywhere.'"

After prebreathing was complete, the crew started pressure checking the suit, lowering the pressure in the airlock while the suit pressure regulator maintained 4.3 psi. Once it was demonstrated that suits were maintaining pressure properly, the hatch was opened and the spacewalkers went outside.

Peterson said that during the spacewalk his suit leaked pretty badly for about twenty seconds and then stopped. The ground didn't know about the leak at the time or they would have stopped the EVA, he said. "I was working with a ratchet wrench. We were just testing tools and stuff. We had launched a satellite out of a big collar that's mounted in the back of the orbiter, and the collar . . . had to be tilted back down before we could close the payload bay doors and come home. So instead of driving it with the electric motors, they said, 'Let's go back and see if we can crank it down with a wrench, to simulate a failure. Suppose it failed, and we'll see if we can do that.'"

Peterson chose not to use the foot restraints provided to help hold him in place, believing it took too long to set them up and move them around.

So I just held on with one hand, actually, to a piece of sheet metal, which is not the best way to hold on, and cranked the wrench with my other hand, and my legs floated out behind me. So as I cranked, my legs were flailing back and forth, like a swimmer, to react to the load on the wrench. The waist ring was rotating back and forth, and the seal in the waist ring popped out, and the suit leaked bad enough to set off the alarms. We did not know what it was. I stopped and said, "I've got an alarm." Story stopped what he was doing and came over.

The seal popped back in place and the leak stopped, and the astronauts finished the EVA.

"In those days we didn't have constant contact with the ground," Peterson added. "They didn't see that. They weren't watching at the time that that happened. They didn't have any way to watch. By the time we dumped the data from the computer to the ground that showed that leak, we were already back inside the orbiter. Then they called up, and they were all upset about what happened here and what was that. We said, 'Well, we really don't know. We got an alarm. The alarm stayed on for about twenty seconds or so, and then it went off, and everything seemed okay. So we just finished what we were doing.'"

At the time, it wasn't known what caused the alarm. The best guess after the mission was that Peterson had been working so hard that he had been breathing more heavily, forcing a higher oxygen feed level and setting off the alarm, an explanation Peterson found dubious. It wasn't until two years later when a similar thing happened to astronaut Shannon Lucid during an EVA training exercise that NASA figured out what really happened. Lucid was in her suit in a vacuum chamber walking on a treadmill when an alarm in her suit alerted that the oxygen flow rate was too high.

"That means that you're pumping oxygen from the tank into the suit, but that also means the oxygen is going somewhere," explained Peterson. "It's going out of the suit somewhere. So they knew they had a leak in her case, and they could also see the oxygen coming into the vacuum chamber, because they were getting pressure inside the chamber."

Lucid stopped walking, and when she stood still, the leak stopped. A technician there recalled that something similar had happened on STS-6. "They went back and got the video of my flight and looked at it," said Peterson.

He said—and this is kind of interesting—when Shannon Lucid was walking, since she's a woman, her hips swivel, and her suit was actually rotating, and we'd never seen that with a guy because guys don't walk that way. But he said, "That's the same thing that happened to Peterson's suit two years ago." So then they went in and changed the seals and all and fixed the problem. But it always amazed me that those guys were dedicated enough to have that kind of memory fixed in their heads. . . . Of course, I got a lot of insulting calls from that guy, "You know, your hips move just like Shannon's." I said, "Not for you."

The EVA afforded the two spacewalkers an amazing opportunity to do some stargazing. While the Space Shuttle normally flies with the payload bay toward Earth all the time, Peterson and Musgrave thought it would be neat to look out at the night sky during one of the passes over the dark side of Earth. "We went to the flight control team and said, 'Guys, when we get on the dark side, we'd like you to roll the vehicle over so we can look out,'" recalled Peterson.

Pete Frank said, "Oh, just for you guys' amusement, you want us to roll the damned vehicle upside down?" And we said, "Yes, you know, wouldn't that be great?" So what they did was even better than that. When they were on the daylight side at noon, they went into what I called the Ferris wheel mode. . . . We went around the Earth holding one [orientation, relative to the sun]. So we went around the Earth so that when we got on the dark side, we were faced exactly away. But because they did that, with the cameras running and all, we got some beautiful pictures of the Earth from a lot of different attitudes that we wouldn't have gotten otherwise. So we got on the dark side, and Paul Weitz, the commander, said, "Okay, guys, you asked for this. Now stop whatever the hell you're doing and look." So we did, and there's lot of light in the payload bay, and the helmet's got these big things. You couldn't see anything. I mean, it was just too much glare. So we got over in one corner and kind of shielded our eyes, and you could see a little patch of sky, but that was about the best we could do.

Peterson was surprised that if he was in a place where sunlight shone into the helmet, he could feel the sunlight on his face. "The visor protects you from the ultraviolet and all that, but you could feel the heat as soon as the sun came in through the visor."

While the views were interesting, Peterson described the EVA spacesuit as extremely stiff and said the gloves were hard to work in.

EVA would be fun if the suits weren't so hard to work in. The suits are fairly uncomfortable. . . . They're not pressurized like an automobile tire, but they're pressurized so they're fairly rigid. The suit has a neutral position. If you just blow the suit up and nobody is in it, it goes to a certain position. If you move it away from that position, it tries to come back because the arms and all are very rigid and they're under pressure. So anytime you're doing anything in the suit, you're typically fighting against the suit itself. The gloves are the same way. If you look at a lot of photography from spacewalks, you see people don't grab something like [they would on Earth], because to do that, you've got to fight the glove. They wedge things between their fingers, and that way you don't have to exert pressure.

The fit of the suit was very important. Because the body expands a little bit in zero g, the fit of the suit has to allow for a certain amount of expansion, and they really can't replicate that expansion on the ground, meaning that the suits generally fit tightly. "When I stood up in it, I could plant my heels against the heels of the boot, and the shoulder harness was right against my shoulders, and the top of my head was right against the top of the helmet," recalled Peterson.

The gloves have to be really close to your fingertips. If they're more than about an eighth of an inch off, you'll lose your ability to feel things and to do precise movements. The problems we had had was, at least in some of the early programs, the gloves were too tight on the fingertips. What they'd do is they'd pinch your fingernail. Several guys lost their fingernails; not while they were in orbit, but it pinched them so bad that it pulled the roots loose in the back. It's very uncomfortable. I mean, it used to be a form [of] medieval torture once to hurt people's fingernails.

STS-7
Crew: Commander Bob Crippen, Pilot Rick Hauck, Mission Specialists John Fabian, Sally Ride, and Norm Thagard
Orbiter: *Challenger*
Launched: 18 June 1983
Landed: 24 June 1983
Mission: Deployment of two communications satellites

With the sixth shuttle mission completed, Bob Crippen, pilot of the first shuttle mission, was about to become the first astronaut to fly on the vehicle twice, this time as commander of STS-7. Historically, the commander had been the most prominent member of the crew, but that was certainly not the case with STS-7, a fate that Crippen had a hand in bringing upon himself. "I essentially helped pick the crew, with John [Young] and George [Abbey], so I would say I had a great deal of influence. And yes, the crew was 'Sally Ride and the others,' which was just right for us."

Before Ride could become the first female NASA astronaut to fly in space, two important decisions had to be made. First was the decision as to whether the time was right for the historic move of including a woman on a NASA spacecraft crew. That decision, Crippen said, was an easy one. The flight would be the first to be crewed by members of the TFNG class of astronauts. Since, of the twenty-one mission specialists in the class, six were female, it seemed appropriate to Crippen and his superiors that one of them should be chosen for the crew.

That decision was followed by another: who would be chosen for a place in the history books. "They were all good, and any of them could have been the first one," Crippen said.

I thought Sally was the right person for that flight for a number of reasons. She was one of our experts on the remote manipulator system, which was critical to what we were doing on this mission. I liked her demeanor, the way she behaved. She fit right in with everybody, as all of them did, but we just got along well, and I thought that's really important when you've got a crew, because you've got to work together. I knew that she would integrate well with the other crew members that we were going to have on board, which initially was just going to be Rick Hauck and John Fabian and myself. We later added Norm Thagard to that flight as well. But she was just the right person to do it at the time.

Ride recalled being informed of her selection to the crew and the distinction of being the first American woman in space.

I met alone with Mr. Abbey, which is a little bit unusual. The commander is the first to know about a flight assignment; Bob Crippen, who would be the commander of my crew, had already been told. But then usually the rest of the crew is told together; at least that was the way that it was done then. But in this case, Mr. Abbey told me first, before he called over the other members of the crew. After I met with him, he took me up to Dr. Kraft's office and Dr. Kraft talked with me about the implications of being the first woman. He reminded me that I would get a lot of press attention and asked if I was ready for that. His message was just, "Let us know when you need help; we're here to support you in any way and can offer whatever help you need." It was a very reassuring message, coming from the head of the space center. [My family] were pretty excited. They knew that this was something that I'd wanted to do for a long time. After all, I'd been in training for four years when I heard the news, so they'd been preparing for this eventuality for four years. They were really excited when I got assigned to a flight.

Ride said NASA did a good job protecting her and the rest of the crew from too much media attention prior to the mission so the astronauts could focus on their mission objectives. "I did very few interviews from the time that we entered training until our crew press conference and the interviews afterward," recalled Ride.

Then we did no more interviews until our preflight press conference about a month before the flight. Right after that press conference, we did a day of solid interviews. NASA protected me while we were in training, and even the day that we did all our interviews, we did them in pairs. I did most of my interviews with Rick Hauck or Bob Crippen. NASA's attitude was, "She's going to get all the attention, and we need to help her." And they did. They did a really good job shielding me from the media so that I could train with the rest of the crew and not be singled out. We also tried to get across that spaceflight really is a team thing.

The training leading up to the flight was particularly intense. Only the commander, Bob Crippen, had flown before, which meant the four rookies had a lot to learn. "The training really accelerated and intensified during that two months before the flight," recalled Ride. "I was spending virtually

20. Sally Ride on the flight deck during STS-7. Courtesy NASA.

all my time trying to learn things, what I'd learned, practice and just stuff that one last fact into my brain. I was barely watching the news at night and really wasn't aware of all the attention. Of course, I was a little bit aware of it—I couldn't help but be—but it wasn't impacting my training at all."

Ride's inclusion on STS-7 created learning opportunities for ground support teams accustomed to all-male astronaut crews. By that point, the JSC team had already been through any number of big-picture changes, such as the locker-room question. Now they were discovering all the little changes that went along with integrating shuttle crews. For example, every item astronauts would use in space had to be chosen and reviewed before flight. "The engineers at NASA, in their infinite wisdom, decided that women astronauts would want makeup," Ride recalled.

So they designed a makeup kit. A makeup kit brought to you by NASA *engineers. You can just imagine the discussions amongst the predominantly male engineers about what should go in a makeup kit. So they came to me, figuring that I could give them advice. It was about the last thing in the world that I wanted to be spending my time in training on. So I didn't spend much time on it at all. But there were a couple of other female astronauts who were given the job of determining what should go in the makeup kit and how many tampons should fly as part of a flight kit. I remember the engineers trying to decide how many tampons should fly on a one-week flight. They asked, "Is one hundred the right number?" "No. That would not be the right number." They said, "Well, we want to be safe." I said, "Well, you can cut that in half with no problem at all." And there were probably some other, similar sorts of issues, just because they had never thought about what just kind of personal equipment a female astronaut would take. They knew that a man might want a shaving kit, but they didn't know what a woman would carry.*

Astronaut Rick Hauck recalled that Ride was very professional, very industrious, was always thinking about the objectives for the flight, and was also a good teammate with a good sense of humor. Reporters at press conferences would focus most of their inquiries on the first flight of a U.S. woman, and that was just fine with Hauck. "I remember one press conference just before we flew," he said. "Someone from *Time* magazine or something said, 'Sally, do you think you'll cry when you're on orbit?' And of course, she kind of gave him this 'You got to be kiddin' me' kind of look and said, 'Why doesn't anyone ever ask Rick those questions?'"

Norm Thagard was a late addition to the crew, chosen in order to address a particular issue that had been plaguing NASA crews since *Apollo 9*, and particularly since the Skylab missions in 1973. "NASA was concerned about this 'space adaptation syndrome,' or upchucking in space, and we wanted to learn more about it," Crippen said. "We had some physicians in this thirty-five group, and we figured, 'Well, we've only got four of us on board. There's room for more. Why don't we pick a doc, and let that individual go through and see if they can figure out this problem a little bit more?'"

Thagard was selected for the mission, and he and other physician-astronauts put together a series of experiments to determine the cause of space adaptation syndrome. "He was wanting some of us to participate in the experiments," said Crippen. "When some of [us] weren't really occupied with things, then we planned on going down there and seeing whether Norm could make us sick or not, which he worked very hard at."

Hauck recalled being subjected to one of Thagard's tests while in space.

As soon as we got on orbit, I got down into the middle deck, and Norm had these visual things that I had to watch, and they're spinning, and, boy, I felt miserable. I mean, they accomplished their purpose. At one point I said, "Hey, guys, I've had it. I'm going to go into the airlock," which was just a nice place to go hide. And I said, "I'm going to close my eyes, and please don't bother me until I come out." I didn't know whether I was going to throw up. I just felt miserable. And I guess it was about four hours later I started to come out of that and that resolved itself.

While STS-7 was best known among the American public for the inclusion of Ride in the crew, the primary mission objectives were a bit more practical. STS-7 would be the first shuttle flight to do "proximity operations," to rendezvous with and capture another object in space. A Shuttle Pallet Satellite in the cargo bay of the shuttle was picked up with the remote manipulator system, taken out of the bay, released into space, and then recaptured.

Ride used the robotic arm to lift the small satellite out of the cargo bay and release it. Crippen then flew the shuttle approximately one thousand feet away from the satellite, circled around it, and rendezvoused, and Ride recaptured it with the robotic arm. "We were to evaluate how easy [or] how hard it was to maintain formation with the satellite," said Hauck, who piloted the shuttle for the second set of rendezvous tests after Crippen flew the first set. "It was just wonderful to be the new guy on the block and be given this responsibility to do some of the flying around this satellite. It all went well, and we got a lot of good data and wrote up a good report on it."

The proximity operations led to another historic moment for the Space Shuttle program: the first photo of the orbiter in space. Today, pictures of the orbiter in space are commonplace; the vehicle was imaged extensively during its rendezvous with the International Space Station. At the time of STS-7, however, that particular sight had never been seen. "While we were headed out, a guy by the name of Bill Green that works up at headquarters had come up with the idea of putting a camera on board this Shuttle Pallet Satellite," Crippen said. "So we actually remotely took pictures of the shuttle from the satellite when it was about a thousand feet away, which gave us some unique shots when we returned back."

Images of the orbiter in space were scripted into the mission objectives, but what wasn't officially planned was an iconic shot of the orbiter with the robotic arm in the shape of the number 7. "On our mission patch we had had the orbiter with the arm up in the shape of a seven, so we concluded that, hey, we might as well do that," recalled Crippen. "We had practiced this on the ground by ourselves, and I think it was Sally who put the arm in that configuration. When we took the picture, it almost looked like the patch. However, Mission Control had not seen the arm go into this particular position before, and they were worried that we were getting it into some limits that it shouldn't be in. It wasn't, but it did cause some consternation on the ground, I think."

Fabian recalled planning the picture with Ride and Crippen.

We really worked hard on that, and we got a lot of help. We worked out the position, the arm in the shape of a seven for the seventh flight. And we didn't tell anybody about this, of course. We had this on kind of a back-of-our-hand type of procedure, what angles each joint had to be in order for it to look like that. And then we had worked on the timing so that we could catch the Space Shuttle against that black sky with the horizon down below. That was the picture we most wanted. We most wanted the shuttle against the black sky and the Earth's horizon down below. I was real proud of that. I was real proud of the work that Sally and I and Crip had done on getting that ready to go. Because it gave you a really strong indication that, you know, this is a spaceship we're talking about here.

The mission also included the deployment of two communications satellites. Once in space, the astronauts opened the cargo bay doors and secured the Payload Assist Modules, which were the boosters used to move the satellites from the shuttle's orbit to their higher destination orbit. "It was really important work, and we screwed it up a little bit," Fabian said. "We threw the heater switches out of sequence, which in the simulations that we did would not have meant anything. But it turns out that those switches had been rewired to do a secondary function. It had nothing to do with the heaters. . . . It never got picked up by the trainers. It never got picked up by anyone that was associated with preparing us to go fly that there

21. *Challenger* in orbit on STS-7 with the remote manipulator system arm in the shape of the number 7. Courtesy NASA.

was something quite different about these switches and that if you throw them at the wrong time, then something unexpected is going to happen."

Unbeknownst to the astronauts, flipping the switches in the wrong order caused the early extraction of several pins on the rotation table that were keeping the satellites from rotating prior to their deployment. "We didn't know anything at all was wrong until the ground told us that we had inadvertently pulled the pins and that they were trying to find a work-around," Fabian said. "They found a way to command from the ground to put the pins back in, fortunately. But this was one of those things; this is a very complex machine, and in spite of everybody's best intentions, sometimes some things slip through the cracks. We had been through this thing in the simulator dozens, if not hundreds, of times, doing it precisely the way that we did it in orbit without it ever coming to anyone's attention."

With his promotion to commander on STS-7, Crippen was able to accomplish a goal that he had set for the position during STS-1: "I was the commander now, so I was going to get to look out the window more."

In fact, Crippen encouraged his crew members, on STS-7 and all of his flights, to take advantage of the unique opportunities spaceflight offers as much as possible. "Enjoy it," he said. "Enjoy it. You never know whether you're going to get a second flight or whatever, so take advantage of when you go up, to really savor it. I had ended up with four flights, and I still remember them today, and every bit of it was enjoyable."

During the mission, Ride talked to CapCom Mary Cleave, the first time a woman on the ground talked to a woman in space. Ride said it didn't even cross her mind. "I don't know whether it occurred to Mary either. I didn't even think about it until after I landed and somebody pointed it out to me. And apparently the first time we talked we said something totally unmemorable. I don't know what it was, but it was not particularly historic."

Cleave's personal impression of the exchange from the ground was similar to Ride's orbital experience, but she didn't have to wait until after the flight to have it pointed out. "I didn't even notice it. Here's Sally and I, we didn't even notice it. But I was on duty, and one female reporter, who will go unnamed, afterwards said, 'Mary, it was so disappointing.' And I said, 'What do you mean?' 'You and Sally just had a normal conversation.' 'Yeah. We were working.' 'Well, you should have said something special for this momentous occasion.' 'What momentous—?' 'First female-to-female communication.' And I went, 'Oh, we didn't notice.' Sort of like, 'Well, I'm sorry I disappointed you, but really we didn't notice it.'"

One memorable event that occurred during the mission made a very literal impact. "We were hit on the windshield by a small fleck of paint, and it made a hole about the size of a lima bean halfway through the outer of two panes of glass," recalled Fabian.

We found out after the flight that it was one [layer] of paint the size of a pinhead that had hit us. And it hits you with such enormous velocity that the kinetic energy associated with that small fleck of paint is enough to blast out that kind of a crater. Crippen decided not to tell the ground that we'd been hit, and it didn't come up until after the flight. And his rationale for that, I assume, was that there wasn't anything that the ground could do to help us. The event had already occurred. We were perfectly safe. They would worry a lot. And so he elected not to say anything. I think after the flight someone said something about that, but I think it was the right decision.

Finally, the time for landing came, and went, and came again. STS-7 was to be the first shuttle flight to land at Kennedy Space Center, but weather was an issue. "John [Young] was the weather checker in our Shuttle Training Aircraft at Kennedy, and he saw some weather, which it can develop very rapidly, which he and I both know," Crippen recalled. "He properly waved us off on that first time, and we elected to wait another day and try it. . . . But truthfully, that extra day we got on orbit was a free day. There wasn't any real work to be done, so all of us had an opportunity to sit back and enjoy, and maybe play a little bit while we were on orbit."

To fill the time the astronauts held the first-ever Olympics in space. According to Hauck, the crew was looking for ideas to kill time. "I forget which one of us said it—because this was near the Olympics—'Let's have a Space Olympics.' Crip was a little wary of this, but he said, 'Okay, what do you want to do?'"

The rules were set. Each astronaut would start on the mid-deck, go through the portside access to the flight deck, across the flight deck to the starboard entryway, down and across the mid-deck, and back up to the top rung of the ladder to the flight deck. Speed was the goal.

"We gave out five awards," Hauck said. "Sally won the fastest woman. John Fabian won the competitor that caused the most injuries. No one got hurt, but I think his leg hit Crip coming around at one point. I think Norm Thagard was the fastest man. Crip was the most injured, and I was the most something. I don't know what it was."

The next day it was time to try the landing again. Crippen's promotion from pilot to commander meant that this time, as commander, he actually got to pilot, but his opportunity to make the first shuttle landing at Kennedy wasn't to be—bad weather in Florida continued that day as well, and the landing was moved to Edwards. The shift scuttled some planned encounters—not only the astronauts' reunions with family members waiting in Florida, but also a chance to meet President Ronald Reagan, who had planned to be on hand to welcome the crew home.

After landing, Crippen said, came the satisfaction of a job well done. "Anytime you work hard to do something and it comes out well, you can't help but feel good, and that's the way I felt with this. I felt my crew had done a superb job. We had accomplished all the mission objectives. We'd made the proximity operations look good. We had deployed a couple of communication satellites, and everything worked. So that gives you a nice high."

After the flight, the attention returned to America's first flown female astronaut. "I think that's when all the attention really hit me," Ride said.

While I was in training, I had been protected from it all. I had the world's best excuse: "I've got to train, because I have this job to do." NASA was very, very supportive of that. So my training wasn't affected at all. But the moment we landed, that protective shield was gone. I came face-to-face with a flurry of media activity. There was a lot more attention on us than there was on previous crews, probably even more than the STS-1 crew. I'd done my share of public appearances and speeches before I'd gone into training, so I knew how to talk to the press and I knew how to go and show my slides and give a good speech. But just the sheer volume of it was something that was completely different for me, and people reacted much differently to me after my flight than they did before my flight. Everybody wanted a piece of me after the flight.

As commander of the mission, Crippen said he felt a responsibility to shield Ride from any unwanted attention. "She was a big hero as we went around, and everybody wanted to meet Sally. There were so many people trying to get after her or get to her, for whatever reason, that part of the commander's job was to make sure that she was protected from that, without being overprotective; just whatever she wanted to make sure that she didn't get overwhelmed by it."

STS-8
Crew: Commander Dick Truly, Pilot Dan Brandenstein, Mission Specialists Dale Gardner, Guion Bluford, and Bill Thornton
Orbiter: *Challenger*
Launched: 30 August 1983
Landed: 5 September 1983
Mission: Launch of Indian satellite, microgravity research

While STS-7 marked the first flight of a female American astronaut, STS-8 broadened the diversity of the astronaut corps further, with the launch of Guy Bluford, NASA's first African American astronaut to fly. As with Ride's flight, there was interest from the media and the public in the first African

American in space. Bluford became a role model for young African Americans. On the twentieth anniversary of his first flight, Bluford said in a NASA interview that it had never been his goal to be the first African American in space. "I recognized the importance of it, but I didn't want to be a distraction for my crew. We were all contributing to history and to our continued exploration of space. I felt I had to do the best job I could for people like the Tuskegee Airmen, who paved the way for me, but also to give other people the opportunity to follow in my footsteps."

Again, as with Ride, NASA protected Bluford from too much media attention before the flight so he could focus on preparing for the mission. He was also spared somewhat because there was a lot of attention still on Ride, whose historic flight had occurred just two months prior.

The crew started out as a crew of four, but approximately five months before the flight Bill Thornton joined the crew to continue studies on space adaptation syndrome, just as Thagard had done on STS-7.

The primary task of the mission was the launch of the Indian INSAT-1B communications and weather satellite. "We were launching a satellite for India," explained mission pilot Dan Brandenstein, "and to get it in the proper place, you kind of worked the problem backwards. Okay, they want the satellite up here, so then you've got to back down all your orbital mechanics and everything, and basically it meant we had to launch at night. The fact we launched at night meant that we would end up landing at night."

This would be first time the shuttle would launch and land in the dark. Bluford said the crew trained for the nighttime launch and landing by turning out the lights in the Space Shuttle simulator. "We learned to set our light levels low enough in the cockpit so that we could maintain our night vision, and I had a special lamp mounted on the back of my seat so that I could read the checklist in the dark," recalled Bluford. "The only thing that wasn't simulated in our launch simulations was the lighting associated with the solid rocket boosters' ignition and the lighting associated with the firing of the pyros for SRB and external tank separation. No one seemed to notice this omission until after we flew."

That omission set the astronauts up for quite the surprise during the actual launch. "We had darkened the cockpit to prepare for liftoff; however, when the SRBS ignited, they turned night into day inside the cockpit.

22. Guion Bluford exercises on the treadmill during STS-8. Courtesy NASA.

Whatever night vision we had hoped to maintain we lost right away at liftoff," said Bluford.

Space Shuttle *Challenger* launched in the wee hours of the morning on 30 August 1983. "We were crossing Africa when I saw my first sunrise on orbit," Brandenstein reflected, "and to this day, that is the 'Wow!' of my spaceflight career. Sunrises and sunsets from orbit are just phenomenal, and obviously the first one just knocked my socks off. It's just so different. It happens relatively quickly because you're going so fast, and you just get this vivid spectrum forming at the horizon. When the sun finally pops up, it's just so bright. It's not attenuated by smog, clouds, or anything. It's really quite something."

Bluford also had fond memories of that first sunrise in space. "I still remember seeing the African coast and the Sahara Desert coming up over the horizon. It was a beautiful sight. Once we completed our OMS burns, I unstrapped from my seat and started floating on the top of the cockpit. I remember saying to myself, 'Oh, my goodness, zero g.' And like all the other astronauts before me, I fumbled around in zero g for quite a while before I got my space legs."

It proved to be a good thing the sunrise was quick because the astronauts didn't have time to enjoy the scenery. "After you're on orbit, you're floating around and that's neat, and you're getting to see the view and that's neat, but after you get up, you've got an awful lot to do in a very short time, getting the vehicle prepared to operate on orbit," Brandenstein said. "There are checkpoints. If you don't get things done or something doesn't work right, you have to turn right around and come back, so you're pretty much focused for about the first four hours up there, of getting that all done. Once that was done, well, then you look out the window a little bit more. I remember when the real work of the day was pretty much over and it was time to go to sleep, you didn't. You looked out the window."

8. The Next Steps

With each mission, the Space Transportation System continued to expand its operational functionality. The first four flights had demonstrated the system's basic capabilities, and the next four had revealed its capabilities as a launch vehicle. After a brief change of pace on STS-9, during which the orbiter served as a space-based science laboratory, the shuttle resumed the expansion of its capabilities as a payload launch system.

STS-41B
Crew: Commander Vance Brand, Pilot Robert "Hoot" Gibson, Mission Specialists Bruce McCandless, Ron McNair, and Bob Stewart
Orbiter: *Challenger*
Launched: 3 February 1984
Landed: 11 February 1984
Mission: Launch of two communication satellites, first flight of the Manned Maneuvering Unit

The first nine Space Shuttle missions received relatively straightforward designations. Each was given a number, and that number was the order in which they flew. But beginning with STS-10 (which was cancelled and later flew as STS-51C) and STS-13, the agency decided there were problems with naming missions this way.

Bob Crippen was commander of the mission that would have been STS-13. "My friend Jim Beggs, who was the administrator of NASA, had triskaidekaphobia (fear of the number 13), and he said, 'There's not going to be [another disaster like] *Apollo 13*, or a Shuttle 13, so come up with a new numbering system.'"

Astronaut Rick Hauck had a similar recollection, though he remembered the decision also being partly inspired by a desire to avoid confusion down the road. "It's my sense that there was someone that decided, 'We are not going to fly a mission called STS-13.' Thirteen-phobic. So at some point, they said, 'Okay, we're going to rename these missions.' And [it was] also because you'd plan a mission, you'd get everybody started on it, and then something would cause that mission to slip past another mission, so that, in itself, was causing confusion. 'We're going to fly STS-12 before we fly STS-11.' So it's easier if you don't number them sequentially."

The new system combined the last number of the fiscal year in which the mission was scheduled plus either a 1 or a 2 for the launch location—1 for Kennedy Space Center and 2 for the planned launch facilities at Vandenberg Air Force Base—plus a letter to designate the planned order. This meant that the tenth Space Shuttle mission became 41B: 4 for 1984, 1 because it was launching from Kennedy, and B because it would be the second mission that year.

Continuing the series of incremental firsts in the early shuttle program, 41B would mark the first use of the Manned Maneuvering Unit (MMU) "backpack" developed by the Martin Marietta Corporation. "It was supposed to be an early-day Buck Rogers flying belt, if you know what I mean, except it didn't have the person zooming real fast," recalled 41B commander Vance Brand. "It was a huge device on your back that was very well designed and redundant so that it was very safe, but it moved along at about one to two or three miles per hour. It used cold nitrogen gas coming out in spurts to thrust you around. The trick was not to let the EVA crewmen get too far out such that orbital mechanics would take over and separate us. We didn't want them lost in space. We didn't want to come back and face their wives if we lost either one of them up there."

When McCandless first began using the MMU, he encountered a couple of problems. First was a slight offset in his center of mass. For the MMU to work properly, the thrust had to be delivered based on the center of mass of the MMU, its wearer, and the spacesuit. As long as those were properly aligned, it would move properly. However, after McCandless found the MMU was not maneuvering quite the way he expected, it was discovered that in the microgravity environment a small offset, such as additional equipment being worn on one side that wasn't factored in, could cause the MMU to move in unanticipated ways.

The other problem McCandless ran into was that the internal thermal control system for the spacesuit was configured assuming that the astronaut would be exerting effort that would cause him or her to generate heat. The idea was that an astronaut would be working up a sweat and the spacesuit

23. Bruce McCandless using the Manned Maneuvering Unit, or MMU, to fly freely in space. Courtesy NASA.

would keep him or her comfortably cool. However, in reality, McCandless found himself cooled far past the point of comfort—it turned out that using the MMU required far less effort than moving from place to place on a regular spacewalk and it resulted in a cooler environment.

"At one point I was actually shivering and my teeth were chattering, and that tends to detract from your overall performance," McCandless said. "If you look at the front of the MMU you'll see a big knob and some markings on the beta cloth running from C to H, which I naively thought meant cold and hot. It turns out that hot just means minimal cooling, and it was set up for somebody who was physically active, that is, with a reasonable metabolic workload in a warm environment such as the payload bay, meaning something that was reflecting heat back and that nominally was maybe 0 [degrees] Fahrenheit but certainly not minus 190 like the effective temperature of deep space."

Initially, when McCandless was maneuvering in the orbiter's payload bay, he didn't have problems. But once he got out away from the shuttle and wasn't as physically active, he said, he got quite cold.

At the time the spacewalk rules precluded shutting off the cooling system, out of fears that, once it was turned off, it wouldn't work if it were turned back on. However, McCandless got cold enough that something had to be done, so Mission Control gave him permission to turn off the cooling system and see what happened. "Well, predictably, after about ten to fifteen minutes it got warmer, pleasantly warm, and a little later it was getting hot so that Mission Control's agreement was I turn it back on, and miracle of miracle[s], it just started right up. No trouble. Over the course of the spacewalk I turned it off and then back on maybe three or four times, and every time it stopped perfectly and then started back up perfectly."

During the flight, McCandless became the subject of one of the more famous photographs in spaceflight history, showing him in the distance floating untethered above Earth. McCandless said that, decades later, having seen the picture countless times, it still connected him to the experience.

When I see it, I guess maybe a little bit of a tingle or goose bumps. It is extremely famous, and I think that perhaps one of the attributes of the photo that makes it so popular is that the sun visor is down so you can't really see my face and it could be the face of anyone or the face of "mankind," whatever you want to call it. And in fact, at Space Center Houston—the tourist facility down there outside of JSC—and a couple of other places, they have a full-size blowup of that with the faceplate cut out so people, mostly kids, could climb up a stepladder behind it and put their faces through it and get photographed.

The photographer was 41B pilot Hoot Gibson. "It's customary during a shuttle mission to swap off or change off duties depending on the mission phase," McCandless recalled, "and this particular phase of the mission, Hoot's job was to use the Hasselblad camera to document the EVA. I think it goes without saying that he did a spectacular job both from the technical standpoint and from the quasi or artistic standpoint of composition and selection of scenes."

Astronaut Jerry Ross was the CapCom for each of the MMU tests. "Those lucky guys," Ross said. "I knew that Bruce had waited a long time and worked many, many years here in the office to get a chance to fly, and so I was happy for him and Bob Stewart when they got a chance to go outside and do their thing." Ross, who later went on to complete nine EVAs in orbit himself, said it was fun to talk to the crew during the MMU demonstrations, but he couldn't help but be a little jealous. "You can imagine how envious I was getting, sitting there on the ground and watching all those guys go out there and have fun."

In addition to the test of the MMU, another of the mission objectives was to test out for the first time the shuttle computers' rendezvous software. The shuttle's first rendezvous target was a Mylar balloon, launched out of the shuttle's payload bay and away from the shuttle by a spring, said Brand. "When it got out a little ways, it timed out and it filled with gas. We were watching it go away from the spacecraft, and all of a sudden it exploded. Not that it was any danger to us, because it was away from the ship, and after all, it was only a balloon. It wasn't like a stick of dynamite. But when this balloon exploded, I said, 'It blew up.' And on the ground they were wondering, 'Does that mean it exploded, or does that mean it filled with gas, 'blew up'?' Well, it exploded."

The crew salvaged the test by tracking the biggest balloon fragment with radar and continuing with the test. "It was fun to do those early things," Brand said. "Many things that are done on the shuttle today as very routine things, back then had to be checked out. The rendezvous system was one of them."

While the equipment test portions of the mission were successful, Brand said the satellite deployments didn't work out so well. The crew deployed two satellites that were similar to the ones successfully deployed on STS-5. The PALAPA-B2 was for Indonesia, and the WESTAR-6 was for Western Union. Each of the satellites deployed flawlessly, and both were supposed to wait half an hour to get some distance from the shuttle before starting the solid rocket motor burn that would lift them into geosynchronous orbit, twenty-three thousand miles above Earth. The engines started the burns but after about twenty seconds unexpectedly stopped. According to NASA, the culprit was the failure of the Payload Assist Module-D (PAM-D) rocket motors. With nothing they could do about it, the crew abandoned the satellites in orbit and returned home.

STS-41C

Crew: Commander Bob Crippen, Pilot Dick Scobee, Mission Specialists George "Pinky" Nelson, James "Ox" Van Hoften, and Terry Hart

Orbiter: *Challenger*

Launched: 6 April 1984
Landed: 13 April 1984
Mission: Deployment of the Long Duration Exposure Facility, repair of the Solar Maximum satellite

Shortly after his return from STS-7, Bob Crippen was once again assigned to command *Challenger*, a post he believes he was given to take specific advantage of the experience he had gained on his last flight. "I think what they were trying to do was to build on the experience that I had from doing the proximity operations," theorized Crippen. "The next, 41C, was going to do our first rendezvous. We had a satellite that was disabled that they needed repaired, so it was similar to what I'd done before, only an extension of it. So maybe that's why I got picked to fly it."

The mission of 41C was to deploy the Long Duration Exposure Facility and to capture the Solar Max satellite that had been launched on a Delta launch vehicle in 1980.

According to astronaut Terry Hart, who operated the robotic arm for 41C, the skills Crippen picked up during years of experience in the astronaut corps were obvious from the very beginning, including in some unusual ways. "It's funny," Hart recalled,

I remember the day we posed for our crew picture. You all put your blue suits on and you bring the helmets in or something, and we took maybe twenty pictures, trying to get us all to have the right expressions on our face or whatever. And then the tradition is, you bring them back to the Astronaut Office and then you ask the secretaries to pick which one is best.

So Crippen and Scobee and Pinky and Ox and I are sitting around, looking at all these pictures. In one of them, one of us would be winking or our smile would be crooked or something like that. Every one of us had maybe a 50 percent hit rate on the pictures, having the right expression on our face. And we looked at Crippen. . . . Every photograph had the same expression on Bob Crippen's face. He had it down pat. He knew exactly how to smile.

The mission marked the first time a shuttle flew a direct ascent trajectory, meaning that instead of launching into a low initial orbit and then using the Orbital Maneuvering System to raise it to an altitude of about three hundred kilometers, the shuttle used the OMS during ascent to achieve a high initial apogee and then used it again to round out the orbit at that level.

From the very beginning of the mission, Hart had difficulty adapting to weightlessness. Not one to have ever had issues with motion sickness on Earth, the space sickness caught him by surprise. "I wasn't weightless for more than three minutes and I knew I was in trouble," he said. "I just felt awful, and I was throwing up, mostly just dry heaves, every thirty minutes or so for a day."

Despite feeling so bad, Hart said that he made sure to get on camera once that first day, just so those on the ground would know that he really was there. "I could barely force myself to get out of the corner of the cabin and get up on camera. There were some things I had to do that first day, but they were minimal. I just had to unstow the arm, and I barely made it through that. I really was totally incapacitated for the first day, and I tried the usual drugs that they give you to help, but I had it so bad, nothing helped at all."

That night, Hart was exhausted and fell asleep fast. But his sleep didn't last for long. "I started dreaming, and I dreamt that I was falling, which I was. I was falling. But I had like a visceral reaction to a fear of falling all of a sudden. I remember I was in the blue sleeping bag and I remember reaching to grab something as I came awake, to stop my falling, and I did it with such force that I ripped the bag that I was sleeping in. It was that violent. And I grabbed on to something, and then I realized where I was."

In Hart's opinion, the repair of the Solar Max satellite was the highlight of the mission. The solar observatory satellite had been in orbit since 1980, observing and studying the sun during the peak of the eleven-year solar cycle. But within a few months of its deployment the satellite had started popping fuses, Hart said. "There was a thermal problem, and some of the fuses got too hot. They had derated the fuses and that had caused them to pop, and the fuses were powering the attitude-control electronics on the satellite. So as a result, the satellite was spinning and they couldn't control it. It was pointed at the sun, but it was wobbling so that it was not of any use to the scientists."

The mission was to capture, repair, and re-release the satellite. It was to be the first on-orbit spacecraft repair, and Crippen and his crew were just the ones for the job. They even called themselves the Ace Satellite Repair Company.

"The Ace thing had come along earlier, actually prior to the shuttle flying," Crippen recalled. He and some friends in the corps helped the wife of a fallen pilot move, and the onetime gig turned into a regular service when other friends needed help moving. "We formed the Ace Moving Company, and our motto was, 'We move single women anywhere and husbands out.' It was mostly a social thing, but we started that and then sort of built on it. Prior to our flight—I believe it was STS-5—they also used the Ace Moving Company sign, because they were deploying satellites. So it was sort of an extension of those earlier days to call ourselves the Ace Satellite Repair, because that was our job to go up and repair the satellite."

Although the satellite was built and launched before the shuttle started flying, scientists at Goddard Space Flight Center had the forethought to design Solar Max with future in-space repairs in mind. Attached to the satellite was a grapple fixture that would allow the RMS to grab it.

The main challenge the crew faced in the task, according to Crippen, was that many of the rate gyros on board had been lost, so it was difficult to stabilize the satellite. The repair required astronauts to undo small electrical connectors, which were difficult to handle while wearing large EVA gloves.

"The other aspect was this was going to be a flight using maneuvering units [MMU]," said Crippen. "That had been done once before without a real task, and this one we wanted to have Pinky [Nelson] go out and actually capture this satellite."

Nelson recalled his excitement at learning that he had been selected for the duty:

I remember meeting with Crippen . . . in one of the little conference rooms over in Building 4, where he doled out the assignments. He assigned me the role of flying the MMU, which kind of made my year, because here was a mission with four military pilots on it, and Terry Hart and Ox van Hoften were both mission specialists, engineer types, but they had also both been fighter pilots, and Scobee had flown everything that had wings, and Crip, this was his third flight already on the shuttle. And they decided to let me fly the maneuvering unit. I never asked why. I didn't want them to think about it.

In orbit the EVA didn't go completely smoothly. Nelson was to fly up to the satellite in the MMU and hook on to the satellite to stop it from moving. Then *Challenger* could move in closer and Terry Hart would grab it with the robotic arm. "We had also practiced, in case something went wrong with that, having

24. James van Hoften and George Nelson on a spacewalk to repair the Solar Max satellite. Courtesy NASA.

Terry grab it while it was still rotating, which was a little bit of a challenge, but we had practiced it and thought we potentially could do that," said Crippen.

"Turned out," he continued, "when Pinky flew out about three hundred feet away from the orbiter, and he came up and did his task perfectly to grab this little fixture, it didn't capture. He sort of bounced off. He tried again and bounced off. I think he hit it in all about three times. The satellite was rotating prior to this around its long axis, but then it started to tumble. So we backed Pinky off, and I was worried that we'd spent all this time training for this, and we were just about to lose it."

In all the preparations, Crippen said, the crew hadn't factored in a tumbling satellite, and they did their best to improvise in the unexpected situation. They considered attempting to go ahead and grab it with the robot arm but were concerned the tumbling would cause the satellite to strike the arm and damage something. They tried having Nelson grab one of the solar arrays by hand to try to stabilize it, but that didn't work either. "We flailed around in there for a while, using up lots of gas, and then finally the ground told us they thought maybe they could stabilize it again with a slow rotation, so they asked us to back away."

Crippen said those EVAs were some of the more worrisome moments of his commanding career: the shuttle was moving to keep a safe distance from the tumbling satellite, while Nelson was floating freely, independently of either.

That's where I got the headache. We knew that we had the digital autopilot set up [to keep a clear] area up above the payload bay. We weren't going to impinge a jet on the satellite or anybody else that was in between there, so that really wasn't that much of a problem. But anytime you've got somebody out there free-flying, you don't want to lose them. So the first thing I wanted to do when we decided we couldn't do it with the tumbling thing was to get Pinky back.

Truthfully, at that time I thought we'd lost it. I could see myself spending the next six months in Washington explaining why we didn't grab that satellite. But the ground had a trick up their sleeve that we weren't aware of, the folks at Goddard did, and they were able to stabilize the satellite.

The next day the orbiter came in for another rendezvous, and it was just like how they'd trained in the simulator, said Crippen. "We went up, and Terry did a neat job of grabbing hold of it, so we captured it. Our fuel had gotten pretty low . . . but we had it. So then Ox and Pinky went out and did their thing of repairing the satellite; worked like champs. They did a couple of EVAs, and sure enough, the first day when they came back in and took their gloves off, all the tips of their fingers were bloody from having to go in and do that fine work."

Once the satellite was in their literal grasp, Nelson said, the repair itself was easy. "It was so much easier to work in space than it is on the ground," Nelson said. "It was a piece of cake. It was so much fun riding on the end of the arm, and just being out there was tons of fun."

Hart, who was operating the robotic arm, said it was an incredible moment for him and the crew when they finally successfully grappled the satellite.

It was euphoric. I mean, we really felt that the mission was at risk, which it was, and we were really on a mission that was demonstrating the flexibility and the usefulness of the shuttle to do things like repair. We were afraid that we were disappointing a lot of people—the scientists, of course, wanting to put the science satellite back into service, but all the people at NASA that were showing what the shuttle could do. In reality, we demonstrated even more just the flexibility of human spaceflight, that you can adapt to things that are unexpected, like this pin and the problems that it caused us. So it was a good opportunity to show even better what the shuttle could do.

The other mission objective was the release of the Long Duration Exposure Facility. A project for NASA's Langley Research Center, the facility was a twelve-sided cylinder that hosted fifty-seven individual scientific experiments. Hart deployed it using the robotic arm. The satellite was so large that it was a very tight fit in the orbiter's payload bay, raising concerns that in deployment it could strike the orbiter, damaging the satellite, or worse.

"The concern there was that I was going to get it stuck, then we couldn't close the payload bay doors, and then we couldn't come home, so we had to be careful," said Hart.

It all went pretty well. First I had to lift it out straight, and then the arm did everything it was supposed to do. And then I think I put it back in again just to make sure it would go back in before I lifted it out one more time to deploy it. We left it out on the arm and did some slow maneuvers to verify all the dynamics and all the things that the engineers wanted to understand about lifting heavy objects out of the shuttle. And then we very carefully deployed it. It wasn't detectable at all when I released it. I mean, it was just totally steady, and we very carefully backed away and got some great photographs of it as we backed away.

The 41C mission also carried an IMAX movie camera that was used to shoot footage for the movie *The Dream Is Alive*. Even though he liked that NASA had partnered with IMAX, Crippen said he worried a little about the size of the camera and he didn't want its use to distract focus from the primary mission objectives. "It was a little bit difficult working some of those things out, but it was a great camera."

In addition to obtaining all the footage IMAX requested of the crew, Hart surprised one of the creators of IMAX, Graham Ferguson, with a special, unplanned thirty-second spot. "We had six film canisters, and we had gone through all of them and we had gotten all the shots that they wanted us to get during the mission," recalled Hart.

I figured I had at least thirty seconds left on the last roll. So I'm kind of, "What can I shoot?' I just want to shoot some indoor thing. And we were in the night side, and Crippen said, "Well, the sun's going to come up in about three minutes here." So I quickly put the camera up and focused on the Earth's horizon just as the sun was starting to break through the horizon. And just as it started to glow a little bit, I ran the last thirty seconds off, and you could see the Earth's limb all illuminated and you could see how thin the atmosphere is from that perspective. And just then the sun blossomed on the horizon, and I ran out of film. So in The Dream Is Alive, *which was the feature they put together from our mission and the two that followed us, there's that sequence in there of the sunrise, where Walter Cronkite's saying, "And here's what an orbital sunrise looks like." So it never occurred to them or us, for some reason, to shoot that particular kind of thing, but when we were up there, we knew that was a dramatic event. So as soon as it was coming, we captured it, fortunately.*

Several weeks after returning home from their mission, crew members watched their IMAX footage in a private screening. "The IMAX people were there and they were all smiles," remembered Hart. "They said, 'You're not going to believe what you did there.' And then they showed the raw footage to us, and it was so vivid in our minds, just being five or six weeks from the mission, that it was almost like being there again, because the IMAX fills your entire field of view with the sensation of being in space."

Despite the agency's best efforts to avoid the number thirteen, the mission didn't escape it completely. Hart said that at the same time the crew designed its official 41C patch, it had also created an underground patch with a black cat and the number 13 on it.

We did our coffee mug with the headquarters-approved STS-41C patch on the front of the coffee cup, and on the back of the coffee cup we had the unapproved black cat with "STS-13" on it. . . . As it turned out, two of the missions in front of us, one mission was canceled, and one mission was delayed. So we ended up being the eleventh flight, as it turned out, anyways. But they also moved the date around. Since it was well before the launch, there was nothing forcing the date, but they just moved the date to get away from the Friday the thirteenth thing, because then it turned out we were going to go early. We were going to launch on the sixth of April and land on the twelfth, but we had a problem during our mission that delayed us one day. So we ended up landing on Friday the thirteenth. But we made it.

STS-41G
Crew: Commander Bob Crippen, Pilot Jon McBride, Mission Specialists Kathy Sullivan, Sally Ride, and David Leestma, Payload Specialists Marc Garneau (Canada) and Paul Scully-Power (Australia)
Orbiter: *Challenger*
Launched: 5 October 1984
Landed: 13 October 1984
Mission: Deployment of a satellite, testing of orbital refueling techniques

Just six months later, Bob Crippen once again served as commander, for 41G. The assignment was unusual; Abbey named Crippen as commander of his fourth spaceflight while he was still preparing for his third. Crippen was surprised to be given two overlapping assignments while there were so many others ready to fly, so he asked Abbey why it was being done that way. "He said he wanted to see how fast we could actually turn people around, so who am I to turn down a spaceflight? I said, 'Sure, but, you know, I'm not going to get to spend as much time training with the crew, so I'd like to make sure I've got somebody there, especially for the ascent portion, that knows how I like to fly the missions.'"

Crippen requested Sally Ride be added to the crew for just that purpose. While Crippen was fulfilling his duties as commander of 41C, Ride and the crew started training for 41G. "I was the only one on the crew who had flown before; the rest were first-time flyers," said Ride.

I was the one that had the experience, and I had also flown with Crip before, so I knew how he liked things done and I knew what his habits were. On launch and reentry I knew what he wanted to do, and what he wanted the pilot to do, and what he wanted the flight engineer to do. . . . Part of my job was to say, "This is the way Crip likes to handle this situation or this sort of problem, and this is how he would want us to work." During the first couple of months, I tried to give the rest of the crew some indication of the way that Crip liked to run a flight and run a crew. Then, thankfully, he launched and landed and came and joined us.

While Ride could play the role of commander in terms of planning training activities, she was still a mission specialist astronaut, not a pilot astronaut, meaning she wasn't trained to front-seat the orbiter during simulations. Jon McBride would sit in his seat as pilot and other astronauts or trainers would fill in as guest commanders.

Originally the crew was scheduled to fly on *Columbia*, which was undergoing modifications after its return from the STS-9 science mission. Leestma was the crew's liaison to follow the progress of the modifications. "It was progressing slowly, and there were a lot of tile modifications that had to be done to *Columbia*. There were a lot of upgrades to make it like the newer vehicles. They weren't going as fast [as planned]. There's always money problems. And so my reports coming back were probably a little bit more negative, only because *Columbia*'s not going make our flight time."

NASA realized that *Columbia* wasn't going to be ready in time, so they shuffled orbiters around and 41G was given *Challenger* and, with the new orbiter, a new payload, Shuttle Imaging Radar B, or SIR-B. "That's when it really got down to training and started getting really serious about it," recalled Leestma. "It was nice to know that we had a payload and an orbiter that we were probably going to fly. At that time I think that *Discovery* and *Challenger* were just kind of flip-flopping all the way along, so it was pretty much a two-orbiter fleet at the time."

Ride's return to space received much less attention than her first flight, even though the crew included two female astronauts and a female NASA astronaut, Kathy Sullivan, conducted an EVA for the first time. "It certainly didn't have the media attention that STS-7 did," commented Crippen.

Judy Resnik had also flown prior to that time, so we'd had a couple of women go fly. So I think the media is easily bored if it's not something that's brand-new. The new thing on this was there was going to be a woman do a spacewalk. What was unusual is, as soon as we named her to do it, the Russians put up a woman and had her do a spacewalk just so she could beat Kathy Sullivan, who was going to do ours. But, in recollection, I don't recall having to deal with anything like what we had seen on STS-7 or STS-1.

Leestma, who hadn't experienced anything like those two historic flights for comparison, had a different perspective:

The media made obviously a very big deal about that. Sally, being the very first American woman to fly; Kathy, now, on this flight, is going to become the first American woman to do an EVA. Two women at one time, how is this going to work? They played it down very well. Both of them were very, very good about it. It was just, "Hey, we're just part of the crew." And Jon and I could easily just stand in the background and just be one of the crew. It actually took a lot of the spotlight off of us, which was fine. Since Crip wasn't there most of the time, it was mostly Sally and Kathy being the spokespeople for the crew, which was perfectly okay. [Before launch], Jon McBride and I went off and sat inside and had a Coke and a candy bar and watched Sally, Kathy, and Crip all get interviewed all day long. And we were happy as clams about it; we thought this was great.

Author Henry S. F. Cooper Jr. wrote in *Before Lift-Off* that 41G had "more anomalies, glitches, nits, and malfs [malfunctions] than almost any previous mission. It was reminiscent of a long film." One of those was when a Soviet laser-testing station targeted *Challenger* with a low-power laser that temporarily blinded the crew and caused equipment not to function properly. The incident was alleged to be a "warning shot" in response to the United States' planned Strategic Defense Initiative space-based missile shield and military use of the shuttle and was met by U.S. diplomatic protests.

Other problems, involving no international drama, were still challenges to resolve. Under normal circumstances, said robotic-arm-operator Ride, the Shuttle Imaging Radar (SIR-B) would not have involved the robotic arm. SIR-B was a radar antenna in the payload bay that was unfurled to make Earth observations. It was then folded and stored for the return to Earth. "Because it was a radar, and because it took up a lot of the payload bay, before Kathy and Dave could go out on their spacewalk, we had to fold it back up again," said Ride.

But we had trouble folding it. We couldn't get it to come down all the way to latch. So we had to use the robot arm in a way that it hadn't been intended to be used. We set the arm down on top of one of the leaves of the antenna and pushed down on it, trying to push it down far enough that a latch could grab it and latch it down. If we hadn't been able to do that, the spacewalk might have been cancelled. But it worked quite well. It was pretty easy to push the top piece of the antenna down just far enough to get it to latch. The problem was solved relatively quickly and to everyone's satisfaction, especially Kathy and Dave's. They were afraid they weren't going to get to go out on their spacewalk.

Innovative use of the arm saved the day on another occasion during the mission, too. On the first day in orbit, the mission was to deploy the Earth Radiation Budget Satellite (ERBS). ERBS was designed to investigate how energy from the sun is absorbed and reradiated by Earth. It was to be deployed by the RMS. Ride was to be the primary arm operator and Leestma trained as backup. "We trained a lot together, spent a lot of time in the simulators and going to Canada and doing those kind of things," said Leestma. "It becomes a little bit of a contest of who can do this quicker or better. All those competitive games were played in everything you do."

When it came time to deploy the satellite, Ride let Leestma pull out the arm, do the checkout, and grapple the satellite. She then took over to pull the satellite out of the payload bay, set it up for deployment, and deploy it. Leestma would then stow the arm.

But during deployment, only one of the satellite's two solar arrays would open. "The first solar array went up and we go, 'Okay, we're ready.' And the ground says, 'Okay. Deploy the second one.' We hit the command and nothing happens. Uh-oh. So what do we do? So we wait, and the ground says, 'Well, do the backup command.' So we do the backup, or do it again, whatever it was, and nothing happens and it's just locked in the side. And we're going, 'Oh, no. Now what do we do?,'" Leestma said.

Mission Control told the astronauts to point the satellite toward the sun to see if maybe it was frozen in place, but when they did that nothing happened. The shuttle was about to fly over a dead zone, where they would have approximately twenty minutes of loss of signal with the ground, and the astronauts were plotting how to take advantage of that time. "This was a flight back before we had all the TDRS coverage," Leestma explained, "so we went through long periods of time where we didn't have to talk to the ground or they couldn't see data. We were getting ready to come up over Australia and go through the Canberra site and talk to the ground, and then we would have about a fifteen- to twenty-minute period before we'd talk to anybody, before we'd come up over the States."

Leestma said he and Ride looked at each other and had the same idea: shake it loose.

We said, "Crip, do you mind if we try to shake this thing loose?" And he said, "Go for it. Just don't break it." We go, "Okay. We're not going to break it." So Sally took the arm and goes to the left as fast as she can and stops it and goes back the other way. The rates in the arm are really slow, but it's putting some kind of force into it. She did this once and nothing happened. We did it, I think, twice, and the second time, I went, "Something's moving." So she puts it up to the deployment position and we're watching it, and it slowly moved a little bit, stopped, moved a little bit, stopped, and then it deployed. I went, "Whew!"

Ride quickly positioned the satellite for deployment just as the shuttle came up over the States and back in contact with Mission Control, Leestma explained. "The ground said, 'Okay, we're with you.' And then we said, 'Well, take a look at the satellite. See if we're ready to go.' I don't remember the exact quote, but they came back up and they go, 'What did you guys do?' And we said, 'We aren't going to tell you, but just check it out, make sure that it's ready to deploy.' And they said, 'Everything looks good.' And so we made our deploy time and the satellite worked."

ERBS is one of the longest-running spacecraft missions. It was expected to have a two-year operational life, but the mission provided scientific data about Earth's ozone layer until 2005, more than two decades after deployment. Leestma recalled that improvised decision during the loss of signal as one of the more memorable moments of his spaceflight career.

It's one of those things that you just kind of go, "Whew!" I don't think we ever would have gotten permission to do what we did, except that we just decided to go do it. It was fun. That was an exciting time. And both of us looked at each other and we got these kind of sneaky grins on our faces as we're looking, going, "They would never let us do this, but let's go try it." And Crip let us do it, so that was pretty neat, too. We had a neat crew. The crew was really a lot of fun, because we really melded and meshed well together.

Crippen recalled another highlight of the mission, an experiment to test the shuttle's capability to refuel satellites in orbit, thus extending their lifespan. The test focused on hypergolic fuels, which are dangerous in multiple ways—volatile and toxic. Before the mission, discussions were held as to whether it would be better to perform the test with actual hypergolic fuels or with water, which would simulate many of the properties of the hypergolics without the danger. The plan was to conduct the experiment during an EVA and have Leestma and Sullivan move the hoses, connect them, and transfer the fluid from one tank to another to see if it was feasible.

The commander, Crippen, was very much in the camp of preferring to err on the side of safety, noting that even a relatively minor problem, such as a leak in the payload bay resulting in hydrazine getting on a spacesuit, would mean the crew member could not come back into the cockpit.

Leestma, on the other hand, was one of the people making the argument that the test should be done with hydrazine instead of water. "Hydrazine is very much like water, but it's got a lot of different properties, one of which is that it blows up if it's not handled right. Crip and the safety folks were very, very concerned that we shouldn't do this with hydrazine. We should just do it with water. The heat transfer properties of water and hydrazine are very, very similar, and that's what we really wanted to know."

But while Leestma agreed that using water would provide a great deal of the desired data, he also felt that using water wouldn't really prove that they could do the transfer with hydrazine. In his opinion, the only way to prove you can refuel with hydrazine in space was to actually do it.

Crippen sent Leestma to White Sands Test Facility to learn all he could about the properties of hydrazine and the benefits and risks of doing the test with fuel or water. Leestma spent ten days at White Sands and came back to Crippen with his report. "I came back from there with a real appreciation for the capabilities of this deadly stuff," he said.

Not only does it blow up, but it's really nasty stuff. You can't breathe it. If you get it on your skin, you can get poisoned. So there were lots of concerns that if we do hydrazine, but if it sprung a leak and even got on our spacesuits, how are we going to get back in the airlock? We don't want to bring this stuff back in. We spent lots of time on how much bake-out time we'd have to do, how to get it off our suits. If you get in the sun, can you bake it out so that you don't bring it in the airlock? And then if we do come back in the airlock, how can you test whether you brought any in with you? How do you get rid of it?

However, despite all those concerns, Leestma still believed the right course of action was to test in orbit using the actual fuels, making sure to establish and use effective safety protocols. Leestma reported all the way up to Aaron Cohen, JSC director of engineering, briefing everyone on why the mission should use hydrazine and not water. Cohen signed off on it, and Crippen finally did too. "Crip probably had the final say-so on that, and he agreed to have us do it with hydrazine, because he had watched me several times in the neutral buoyancy facility to do the whole procedure, and how careful we were, and we had triple containment of the liquids at all times."

When it came time to do the EVA, Leestma described feeling like he was on top of the world.

Going out the hatch and getting your entire faceplate filled with this Earth—which is just a spectacular sight—it's emotional and spiritual. The Earth is an incredible creation. Your heart rate goes up and you're going, "I cannot believe I'm doing this." You're going almost eighteen thousand miles an hour and you're weightless. You've got this four-hundred-pound suit on, and yet you can move yourself with just a finger. Your faceplate is filled with all these clouds and ocean and ground and greens. It brings tears to your eyes. I actually had tears going on. And you don't want to have tears in your eyes, because you can't do anything about it inside the faceplate. So you've got all these things going on. "What's going on here? Calm down, Dave. You've got a job to do." But it's a very emotional rush.

Then it was Sullivan's turn to exit the airlock as the first woman to do a spacewalk.

Kathy comes out, and you know what's going through her. You don't really communicate that. And then she's doing something that is historic. Man, you just go, "Wow! Why do I get to do this?" So then you go, "Well, okay. As long as it's me, that's great. Let's press on." It's an emotional thing. And then you're going, at that same time, in the back of your mind, which is always the thought that astronauts have right from the time of launch, is, "Don't screw this up." Make sure you do it by the book and you're very careful and very meticulous, and you're going, "Oh, don't screw this up."

I felt like it was five minutes, and then Crip or Jerry Ross, the CapCom, said, "Dave, it's time to get back in the hatch and wash your hands for supper," or something like that. It didn't last as long as you'd want to.

After weather had shifted his last two landings away from Kennedy Space Center, Crippen, on his fourth mission, finally got the chance to land in Florida. As the orbiter was coming in lower and lower and slowing down as it came, Crippen said he started to notice just how fast the aircraft was flying, something he said he didn't notice as much while on orbit.

I'm not really sure why that is, but it seems like you're going faster, when actually you really are slowing down. I can remember I could see Jacksonville, Florida, when we were over probably in Kansas-Missouri area. I could see the whole peninsula of Florida, and shortly after I picked up Jacksonville, I could see the Cape, because it's very pronounced where it sticks out there where the Kennedy Space Center is. Then there's the shuttle landing facility. Visually I think I picked up everything necessary to fly an entry much earlier than I did while we were coming in to California. Even though we were flying on the autopilot and doing very well, if there had been something wrong with the navigation, I felt like I had the capability to fly it on in and land.

Crippen had practiced a Kennedy landing hundreds of times in the Shuttle Training Aircraft, and the real thing seemed pretty much the same, he said. "I often joke that they've got a fifteen-thousand-foot runway, but they built this moat around it and filled it full of alligators to give you an incentive to stay on the runway. But it worked out well. The landing was fine."

STS-51A
Crew: Commander Rick Hauck, Pilot David Walker, Mission Specialists Anna Fisher, Dale Gardner, and Joe Allen
Orbiter: *Discovery*
Launched: 8 November 1984
Landed: 16 November 1984
Mission: Launch of two communication satellites, retrieval of two communications satellites

"We were scheduled to launch on the seventh of November early in the morning, and we loaded into the spacecraft," 51A mission specialist Joe Allen recalled. "We unloaded from the spacecraft, in spite of the fact that it was a beautiful, crystal-clear day in Florida and in spite of the fact that the equipment seemed to be working properly."

As perfect as both the visible weather and the spacecraft were that day, what the crew could not see was what was going on in the upper atmosphere—high wind shear levels.

Allen said that while the crew of the fourteenth shuttle flight was, of course, disappointed by the delay at the time, he later felt much better about the delay. "History now shows we were also possibly very lucky, because both of the tragic accidents, that of the *Challenger* and that of *Columbia*, involved launching through very high wind shear conditions, and there's some thinking now that high wind shears and Space Shuttles do not safely go together."

The mission was twofold: release two satellites—the Canadian communications satellite TELESAT-H and the defense communications satellite SYNCOM IV-1 (also known as LEASAT-1)—and retrieve the WESTAR-6 and the PALAPA-B2 communication satellites, which were deployed on 41B but failed to achieve proper orbits.

While Allen described the mission as "fairly simple," some members of the crew felt that people underestimated just how challenging the mission would be. "I got several somewhat rude notes from my fellow astronauts underscoring the fact that in delivering the two satellites to orbit and picking two up, that neither Dale nor I was to get these satellites confused," recalled Allen. "In other words, don't bring home satellites that we'd just taken there. It's very unkind from our fellow astronauts to point this out, but it was kind of funny."

Allen recalled meeting several days before the scheduled launch with an associate administrator of NASA. According to Allen, this individual was the newly named head of the Public Affairs Office at NASA. No agenda was given for the meeting, save that it was scheduled to last for about an hour and take place in a meeting room in the crew quarters at Kennedy Space Center.

We went into the meeting room in flight suits, we five crew members, and then in came the associate administrator. Very nice gentleman; introduced himself, and I think one of his aides or deputies was with him. Introduced himself all around. We sat down and Rick [Hauck] said, "Now, Mr. ———"— it's unimportant—"what's the agenda for this meeting?" Whereupon the naïve individual—we didn't know it at the time—said, "Well, no specific agenda. I just wanted to introduce myself and just say that if there's anything I can do for you, I'm here to help and wish you good luck." We as individual crew members were all surprised, because this really was occupying a good chunk of our morn-

ing, and time was very important to us right then. Not that it was discourteous of the individual, but it was unclear why this meeting was to take place.

A day earlier, a story had appeared in the *Florida Today* newspaper in which an unnamed NASA spokesperson quoted the crew as saying that the likelihood of capturing both satellites was very high. The crew members were very curious about the story and, in talking among themselves, determined that none of them had given the quote. It seemed unlikely as well that a member of the support crew would describe the mission in that way. "We were fretful that those words had gone in, because we thought the task was going to be quite difficult," Allen said.

We didn't plan to do anything about it, because newspaper stories are newspaper stories, and it was now water under the bridge. Something about the start of this meeting [with the associate administrator], though, got under Rick's skin as the commander, and he said, "Mr. So-and-so, there is something that you can do." He then cited this newspaper article of two days ago, and he said, "I do not know who said that. I assure you that none of us said that, nor do we believe it. And I will personally tell you that my assessment is, if we successfully capture one satellite, it will be remarkable, and if we get both satellites, it will be a fucking miracle." And he went on to say, "You can quote me on that." Well, the man was shocked, properly so. We were as well. And then Rick said, "If you have no other business, I think this meeting is over." We're ten minutes into a one-hour meeting. He excuses himself, somewhat distressed, with probably good reason, and leaves, and we go back, too, and we sort of said [sarcastically], "Boy, Rick, that was being very commander-like. But good for you."

While his crew members were somewhat surprised by their commander's remarks, Hauck explained that he was livid over the comments in the article.

I thought, "Here we are, NASA is shooting themselves in the foot because we are implying that this is easy." And I had the opportunity to see this gentleman, and I said, "You have set NASA up for a humongous failure by the nature of this press release. . . . In my view, if we get one of these satellites back, it'll be amazing, and if we get both of them back, it'll be a miracle. . . . You have not done NASA any favors." There's no sense in trying to tell the American people and the taxpayers that what you're doing is easy, because it isn't easy. It's very hard, and any implication that it's easy is a disservice to everybody.

Many parties had a vested interest in bringing the satellites back to Earth. NASA was still trying to prove the shuttle's capabilities. The insurance companies that now owned the failed spacecraft wanted them returned. And McDonnell Douglas, who made the solid rocket motors that failed, was interested in studying what happened to cause the motors to fail. "You put those all together, and there was great motivation from all sides to mount a rescue mission to bring those satellites back," said Hauck.

The opportunity to recover the satellites came with a great burden of responsibility, even more than usual, Allen said, because of the insurance companies involved. The two insurance companies were Lloyd's of London and International Technology Underwriters. He explained that while, for insurance companies in the United States, the potential for losses is mostly felt by the corporations, for Lloyd's, one of the first insurers in the world marketplace, there was a much higher level of personal risk for involved stakeholders. "We learned in a rather roundabout way that more than a few lives would be dramatically affected by our success or nonsuccess, lives of people whom we did not know on the ground, but . . . individuals who would lose [a] considerable amount of their personal wealth if the satellites were not recovered."

Allen said he found it bizarre that the insurance industry was now involved in space exploration.

I remember hearing an insurance person from England describe the mission we were about to set off on as very unusual, to his way of thinking, because, he pointed out, he had spent a lifetime insuring things against fire or the chance they would explode, and he said, "With you chaps in the space business, you purposely set fire to a massive amount of explosives, and I find myself now betting on whether you can control the explosion or not." And I thought that was a rather graphic way to describe a rocket launch. The more one thinks about it, though, it's a very accurate way, and sadly, a year later, we saw an example of what happened when we chaps could not control the explosion, and we lost Challenger *as a consequence, as a tragic consequence.*

Capturing the satellites was not an easy assignment, Allen said. Those who built the satellites never envisioned they would be revisited, much less handled by astronauts. The satellites were supposed to be twenty-two thousand miles above the surface of Earth in geosynchronous orbit, far higher than the Space Shuttles could reach, but they had not achieved that orbit.

The only features of the satellites that it might be possible to use as a handle were microwave guides and antennae affixed to the top of the satellites, or possibly the engine nozzle at the bottom of the satellites that was used to reposition them from time to time.

NASA engineers came up with a very clever plan, Allen said. "We decided the antennae would be too fragile to grapple, so they decided we would stab the satellite from the back, using a device that I later called a tribute to Rube Goldberg and Sigmund Freud, the device we called the Stinger."

Allen said the device resembled a folded umbrella. During a spacewalk, the astronauts would put the Stinger inside the rocket and then open it so that the tines of the umbrella would stick against the sides. "When completely suited in the spacesuit and in the MMU with the Stinger device affixed to our chest, we looked for all the world like a space-age medieval knight entering a jousting contest," Allen said.

Allen was looking forward to the opportunity to finally don a spacesuit for the recovery operations, after having missed out on participating in the first shuttle spacewalk back on STS-5, due to a problem with the EVA suit. He noted a bit of irony in the fact that, without the satellite recovery, he likely would not have had a spacewalk on this mission, either—on STS-5 he had lost out on performing a spacewalk because of one malfunction, and now, because of a malfunction on a different mission, he was getting a second chance.

Allen and Dale Gardner were suited up with help from Pilot David Walker. Walker went to place Allen's helmet on his head, but Allen said, "Stop."

I said, "David, stop. I am so hungry. I really need a cookie or something to eat." He said, "Oh, Joe, how could you?" . . . I said, "David, I need a butter cookie." So he goes off into the food pantry, grabs things, throws it hither and yon, and comes back with a butter cookie. I open my mouth—keep in mind I can't use my hands now—he puts the butter cookie into my mouth, the whole thing, and then he hits my jaw shut. He says, "Eat it, but don't choke, you little rodent." I ate the butter cookie; felt better. David put the helmet onto me, popped it. We're now sealed, and there unfolded a pressurizing and then a depressurizing of the airlock, and the EVA started.

Allen maneuvered himself in the MMU to the first satellite and successfully grabbed hold of it. But he was facing the sun and was being blinded by its brightness. He radioed to his crew inside the orbiter, and Hauck moved the orbiter so that a shadow was cast on the satellite and Allen was no longer seeing the sun.

"It was beautiful, clever as could be," said Allen. "I could then see the bull's-eye, the center, the rocket engine, very easily, threaded it like I'd done it all my life, deployed the tines of the Stinger, tightened down the clamp, and voilà, there it was. Anna [Fisher] grappled me with no problem at all; turned around, and Dale set about affixing the clamp to go to the top."

However, the tool designed to fit on the top of the satellite did not fit. The problem, the crew later learned, was that the actual satellite was not built to the drawings the crew had used on the ground.

About the time we realized this was the case, Discovery *came acquisition of signal, and we reported to the ground that the holding tool did not fit, and they said, "Roger that. We'll get the back room working on it. What is your plan?" And we said, "We're going to go to Plan B." "Roger that. See you next AOS," and the ground was now out of earshot again. David Walker, bless his heart, was the keeper of all the Plan Bs that we as a crew, prior to the launch, had devised, and we'd written them down. What would we do in the event that "blank" failed? We had a Plan B for what we would do in the event that this clamp failed, and it was, sad to say, written on David's piece of paper just as "Improvise." We really did not know what we were going to do.*

The crew discussed it and decided that improvisation would involve affixing a foot restraint on the orbiter into which Allen could maneuver the MMU with the satellite. There Fisher would help move the satellite so that Allen could hold the top of it, not with the arm's grapple fixture, but just by getting a handhold. Gardner would then, by himself, attach the large clamp—normally a two-person job—to the bottom of the satellite. Allen and Gardner had successfully maneuvered the clamp with two people in the water tank at Johnson Space Center several times, and even with two people it hadn't been easy.

Allen held the satellite for about two hours while Gardner single-handedly affixed the big clamp. "The fact he was able to do it still astonishes me, but he was," said Allen.

He's just persistent, the most persistent individual I've ever worked with, and one of the smartest, and he did the impossible. Ultimately the clamp was affixed. I later was given far too much credit for supporting a two-thousand-pound satellite for one orbit of the Earth, and a political cartoon appeared in Canadian newspapers the next day showing a chunky little spacesuited crewman standing on the gunnel of an orbiter holding this satellite, and the caption was, "Nobody kicks sand in this man's face anymore," referring back to an old Charles Atlas ad of many, many decades earlier. Dale was not recognized in the paper for the heroic work he had done, but his fellow crew members knew of it and still know of it.

Two days later Allen and Gardner used the exact same procedure to recover the second satellite. Allen recalled,

We were LOS, *loss of signal, and Rick, Anna, and Dave all were very pleased for us and said, "Congratulations," and Rick, as the commander, said, "Joe and Dale, when we come* AOS, *I want you to report that both satellites are locked safely aboard." We looked at each other and kind of shook our heads outside, and almost together, we said, "Rick, that's the commander's job. When we come* AOS, *you report that we have two satellites safely aboard, and you can also use the words 'fucking miracle.'" We came* AOS, *and Rick, in his Chuck Yeager–type relaxed drawl, said, "Houston, Roger.* Discovery *here. We have two satellites safely aboard." You could hear the Mission Control people cheering through the microphone of the CapCom. It was really quite fun.*

Once both satellites were successfully retrieved and stowed in the payload bay, Allen took a photo of Gardner holding a For Sale sign in front of the satellites. "We had prepared in advance of the flight [a sign] that said 'For Sale,' because the satellites would be returned and would then be in the ownership and the possession of insurance companies, which had every intention of selling them as brand-new satellites," Allen explained.

It's a terrific photo, and one of the only photos I've ever taken that shows me as well, because I'm reflected in Dale's helmet, holding my camera, and the photo shows part of the Earth, the blackness of space, Dale Gardner, the For Sale sign, and my likeness reflected in his helmet; a favorite photo of mine to this day.

When we returned, the For Sale photographs—and Rick and Anna and David had taken many from the flight deck as well—were an important part of the press package that went out, and they showed up in a number of magazines. I might say that the Lloyd's of London and the International Technology

25. Dale Gardner holding a For Sale sign after retrieving two malfunctioning satellites during STS-51A. Courtesy NASA.

Underwriters were very, very pleased with these photographs. NASA was not as pleased, and we were given somewhat curt—"reprimands" is the wrong word—curt discussions from our headquarters bosses over what did we have in mind in doing that.

After a nearly eight-day mission, the crew landed with both satellites on board at Kennedy Space Center. Allen said two individuals from the U.S. Customs Department met them in crew quarters.

We were surprised by this, and they said they had forms for us to fill out, because we were bringing into the United States approximately $250 million worth of technical hardware, and there was a certain duty now due on this, because anything that's imported into the United States over a certain value must be taxed, and the tax would be 10 percent of $250 million. How did we plan to pay for that? Fortunately, they also had an agreement between Customs and the NASA Office of General Counsel that waived this import duty, that the chief NASA lawyer, Neil Hosenball, had foreseen as a complication and had organized the waiver prior to the success of the mission. But we were to sign the Customs form, and for that they gave each of us a United States Customs hat.

9. Science on the Shuttle

As the nation's Space Transportation System, the primary goal of the Space Shuttle was just that—to transport people and cargo back and forth between the surface of Earth and orbit around the planet. However, it was quickly obvious that the shuttle had even greater potential.

In 1973 and '74, the Skylab space station had demonstrated the value of conducting scientific research in Earth orbit. From the beginnings of the Space Shuttle program, plans had been that the vehicle would support a space station that would continue, and build further upon, the work done on Skylab. However, the realities of budgetary constraints meant that the two could not be developed at the same time and that the space station would have to wait.

In the meantime, though, the Space Shuttle itself would provide a stopgap measure— a pressurized module placed inside the orbiter's cargo bay could be roomy enough to serve as an effective orbital science laboratory.

STS-9

Crew: Commander John Young, Pilot Brewster Shaw, Mission Specialists Owen Garriott and Bob Parker, Payload Specialists Ulf Merbold (Germany) and Byron K. Lichtenberg

Orbiter: *Columbia*

Launched: 28 November 1983

Landed: 8 December 1983

Mission: First flight of the Spacelab laboratory module

Launched in November 1983, STS-9 marked a new step forward for the Space Shuttle program, while also hearkening back to the first four missions. The first mission to fly the Spacelab science module, STS-9 was both a demonstration and an operational flight. The primary purpose was to make sure that the Spacelab module worked properly, but that was achieved by conducting a full complement of scientific experiments. According to STS-9 mission specialist Owen Garriott, who was a member of NASA's first group of scientist-astronauts and of the second crew of Skylab, the mission of Spacelab I included experiments in biomedicine, astronomy, fluid physics, materials processing, and atmospheric sciences.

A rack of fluid physics equipment enabled the crew to inject liquids, shake them and rotate them, combine them and stretch them, in order to study the small forces associated with the surface tension and the even smaller forces that are associated with fluids that cannot be as easily demonstrated or measured in a one-g environment. An adjacent double rack was dedicated to materials science. The racks included furnaces in which material samples could be heated and resolidified. Other equipment focused light on a crystal, melted it, and then allowed it to solidify more carefully. Spectrometers flown on the mission allowed the crew to conduct stellar astronomy and air-glow research through the orbiter's windows.

The variety of the experiments was what drew Garriott to the flight. Garriott, who spent more than fifty-nine days in space as a science pilot astronaut aboard Skylab, explained that for a mission like Skylab or STS-9 that is focused on interdisciplinary scientific research, the ideal crew member is a scientific generalist, with interest and ability in multiple areas, rather than someone who specializes in one field. "My interest is interdisciplinary work, so I found [the mission] quite interesting, and I think it does relate to the fact that you need an interdisciplinary background to try to conduct experiments for all of these PIs [principal investigators]. You obviously can't have a representative for each PI there, and you need somebody who has got some degree of competency in all of the variety of areas."

Before the flight assignment, during the period between Skylab and shuttle, Garriott had gone into management, becoming director of science and applications at Johnson Space Center. He was still a member of the astronaut corps but was no longer actively involved with the Astronaut Office.

I had my hands full two buildings away. I think I still talked to the folks in the Astronaut Office all the time. I was still flying airplanes. I might have been going over to those weekly meetings as well. But I did not have any acting role in the Astronaut Office from the standpoint of getting ready for shuttle or anything else. I was expecting to stay in the Science and Applications

26. In the Spacelab on the first Spacelab mission are Robert Parker, Byron Lichtenberg, Owen Garriott, and Ulf Merbold. Courtesy NASA.

Directorate until another flight opportunity came along. I'd always intended to return to the Astronaut Office full time as soon as there was the first flight opportunity available.

Garriott started training for the shuttle in 1978, when NASA believed the mission was still only about three years away. "We had a lot of training to do," Garriott recalled, "because there were something like sixty experiments on board, half from Europe, half from the U.S., roughly. And we visited almost every PI at their facilities and talked with them about how to operate it and the hardware development."

During that period the crew was very involved in the development of the Spacelab hardware, particularly the controls and displays the astronauts would use to interface with the equipment. "If you don't understand how [something] works, if it's not something that's human friendly, you can waste a lot of time or make a lot of mistakes," Garriott said. "That was the one thing that supposedly the crew members were more expert in, the interface between man and machine."

Delays meant that Garriott did not make his second spaceflight until a full decade after his first, but he said it was worth it. "That ten-year delay is longer than I expected, but it's a decision that I really made when I went over to Science and Applications in the first place. I wanted to come back and fly again. And so I was just anxious to get back and fly again."

The variety of the experiments meant that the crew members had to spend a substantial amount of time learning about the experiments and working on the ground with the principal investigators with whom they would be interacting from orbit. There had been fewer disciplines represented on Skylab, Garriott recalled, so crew members did not have to travel as much or as far to talk with the principal investigators. "In fact, for most of the time, they came to JSC for our solar physics training and for our biomedical training." On Spacelab, the investigators were all over the world, with experiments from around the United States, from countries in Europe, and one from Japan. "We were very international and traveled all over the world in order to talk with the investigators about the conduct of their experiments. I feel so fortunate to have had the chance to work in all of these different disciplines with the real world's experts and to learn from them."

Garriott also felt fortunate to work with the other astronauts on STS-9, including Commander John Young.

I've always gotten along extremely well with my fellow crewmates. John Young, who was the commander of Spacelab I, is not an Alan Bean [commander of Garriott's Skylab crew]. He's motivated differently, different personality, [a] standard sort of a prototypical test pilot. I was a little concerned about that, since we were bringing in the first foreign international. How well is that going to work? It turns out it worked extremely well, for which I give John Young a lot of credit. I really enjoyed having John as the commander of our flight. And I think after we came back he had no better friend on board our flight than Ulf Merbold. They still often talk. Whenever Ulf comes back, I think they get together for dinner.

While the commander's primary responsibility in orbit was the operation of the vehicle and oversight of the crew, Garriott noted that Young made several other contributions to support the scientific portions of the mission. For example, when the mounting apparatus for a camera used for a vestibular experiment broke, rather than allowing thousands of film frames to go to waste, Young took the camera and film to the flight deck and spent time taking Earth-observation photographs.

"John really jumped in and assisted in the conduct of the science," Garriott commented. "Before flight there was some little concern about that,

that whether or not a standard prototypical test pilot would enjoy and participate. But we all really enjoyed having him on board. He did his share, as did the rest of us."

STS-9 was the first NASA flight to include payload specialists, astronauts who were not career NASA employees. Ulf Merbold, the first international astronaut to fly on the shuttle, was a West German astronaut from the European Space Agency, and Byron Lichtenberg was a researcher from the Massachusetts Institute of Technology (MIT).

Garriott described the feelings of the career astronauts toward the payload specialists as "uncertainty" and compared it to when his class of scientist-astronauts joined the all-pilot astronaut corps.

When we came into the Astronaut Office, I can imagine that the pilots were thinking, "Geez, these old fuzz-hair university types, can they hold their own here, do they know what's going on?" In the same way, these [payload specialists] are really university types here. Are they really motivated to fly, can they hold their own, and so forth? And I very quickly found out that yes, indeed, they could. They were on par with all of us. We were very much of the same kind of breed, in my opinion. We got along extremely well. Everybody got along fine. And after thirty years, they're still some of my best friends. So it was a very pleasant and positive experience for both MS and PS [mission specialists and payload specialists] as far as Spacelab I is concerned. I think that is generally true for most of the flights, though not necessarily in each individual case. When you get up to twenty, thirty people, you're bound to find some rough corners somewhere. I can say on Spacelab I, it was a remarkably positive experience for all of us.

While Garriott did not recall any resentment over payload specialists taking slots that otherwise might have gone to career astronauts, he also noted that in the case of STS-9, the payload specialists had trained for as long as the newest NASA astronauts had been in the corps.

Brewster Shaw, the pilot for STS-9, said that, being in a crew that was mostly rookie space flyers, it was interesting working with veteran astronauts Garriott and Young.

John and Owen were the only two guys who had flown. [Robert] Parker had never flown. I'd never flown, and Ulf and Byron had never flown before. So John and

Owen were the experienced guys, and they kind of were the mentors of the rest of us. It was fun to watch Owen Garriott back in the module, because you could tell right from the beginning he'd been in space before, because he knew exactly how to handle himself, how to keep himself still, how to move without banging all around the other place. And the rest of us, besides he and John, the rest of us were bouncing off the walls until we figured out how to operate. But Owen, it was just like, man, he was here yesterday, you know, and it really had been years and years.

Garriott was among the very small number of astronauts to have the experience of flying on both the Saturn IB rocket and the Space Shuttle. "The launch phase is remarkably similar, because both vehicles have a thrust which is just a little bit more than the weight of the vehicle. You start out at very low acceleration levels," Garriott explained, comparing the g-forces of the vehicles' initial accelerations to what one would experience in a car.

It keeps building up a little bit, but the important thing is it's steady. When you accelerate in a car, you accelerate for three or four seconds and then you reach the speed you want to be at. Here, you keep on accelerating. So you keep up an acceleration which is increasing up to maybe three or four gs. But it takes about eight and a half minutes, nine minutes, to get to space on either one of the two vehicles. As you lift off, there's quite a bit of vibration with either vehicle. Once you get above about one hundred thousand feet, first of all, you stage, you switch to a smaller engine on a second stage, and you're above most of the atmosphere so the vibration diminishes. Then it just becomes a nice steady push, and you continue that push at about three to four gs, all the rest of the way till you're into orbit. All of a sudden you're bouncing up against your straps. So the launch is remarkably the same, in my view, between a Saturn and a shuttle.

With the development, training, and launch complete, it was time for the real business of STS-9 to begin. The crew worked for twenty-four hours a day, divided into two shifts. The commander and pilot each were assigned one shift, and the scientists were split between them. Each crew member was on duty for about twelve hours, with a brief handoff between shifts. The astronauts then had twelve hours off duty, during which to sleep, eat, get ready, and enjoy the experience of being in space.

Shaw remembered that he and Young would spend the majority of their work shifts on the orbiter's flight deck.

You didn't want to leave the vehicle unattended very much, because this is still STS-9, *fairly early in the program. We hadn't worked out all the bugs and everything, and neither John nor I felt too comfortable leaving the flight deck unattended, so we spent most of our time there. We had a few maneuvers to do once in a while with the vehicle, and then the rest of the time you were monitoring systems. After a few days of that, boy, it got pretty boring, quite frankly. You spent a lot of time looking out the window and taking pictures and all that. But there was nobody to talk to, because the other guys were back in the back end in the Spacelab working away, and, you know, you just had this, "Gosh, I wish I had something to do," kind of feeling.*

While he and Young were not heavily involved in the research part of the mission, Shaw did recall being part of one experiment, designed to study how humans adapted to working in space.

I did "Helen's balls." . . . Helen was a principal investigator, and she had a bunch of little yellow balls that were different mass. . . . Since there's no weight, there's only mass, in zero g, we had to try and differentiate between the mass of these balls. You would take a ball in your hand and you would shake it and you would feel the mass of it by the inertia and the momentum of the ball as you would start and stop the motion. Then you'd take another one and you'd try and differentiate between [them], and eventually you'd try and rank [the] order of the balls. . . . And, quite frankly, that's the only experiment I remember doing.

Garriott used his second flight as an opportunity to make history in a way that had a lasting legacy for the space program. When Garriott was younger, his father took an amateur radio class with him, and the two became ham operators. After he was assigned to STS-9, Garriott got permission to take an amateur radio rig with him on *Columbia* and was able to make contact with people on the ground. "I'd say fifteen to twenty hours was reserved for use of the ham equipment . . . after the working day was over," Garriott recalled. "I found that quite interesting and enjoyable, and it turned out to be a very positive thing from the standpoint of the ham radio community."

While Garriott mostly talked with whomever he happened to get in contact with, he had prearranged a few conversations. "There were a few high-profile people that that's the only way you'll be able to get in touch with them. Normally, when you just want to talk to anybody, you use a call signal that's called 'CQ.' And if you call CQ from space, you'll probably get a hundred answers, so you're way overloaded. So if you want to [reach] one particular person, you have to specify a time and frequency that's kept private."

One of the preplanned calls was to Garriott's hometown of Enid, Oklahoma. "I made arrangements to talk with them and my mother . . . with one of the local hams on the radio," he said. "Another one was [Senator] Barry Goldwater, who is a very well-known ham in the Senate and always looked after amateur radio very well from that position. And another internationally known, great enthusiast was King Hussein from Jordan."

For the hundreds of other people Garriott talked to in addition to those preplanned calls, he used a tape recorder to log the contacts. "Everyone I could discern answering my call, I would repeat back their call sign, as many as possible. And then I listened to every one of those tapes. And I had another friend go through the whole list, and we responded to everyone whose call sign we could pick out of there. So there were, I don't know, four or five hundred people that all received a card called a QSL, that's another special symbol that's used by hams as a symbol of a contact between two people."

Garriott's initiative to use amateur radio on STS-9 created a lasting legacy. Almost thirty additional shuttle flights carried ham radio and used it to talk to schoolchildren and other groups, allowing students to ask questions of astronauts in orbit during organized events. Years later, amateur radio became a fixture on the International Space Station, with its first crew conducting a ham radio conversation with a school within weeks of boarding the station. Thousands of students have had the chance to talk to orbiting astronauts as a result of the amateur radio contacts begun by Garriott.

That legacy paid off for Garriott himself in a very special way many years later. In 2008 Garriott's son, Richard, became the first second-generation American space flyer as a spaceflight participant on a Russian Soyuz flight to the International Space Station. While there, Richard was able to use the equipment on the Space Station to talk to his father on the ground. As unique as that experience was, it had an even deeper personal significance for the elder Garriott. "My father and I got our ham licenses together back in 1945, and his call sign was W5KWQ. Mine was W5LNL. Now, normally, when a person dies, they reserve that for a little while, but then they'll eventually reissue the call sign. So after my father died, which would have been back in '81, it hadn't been reissued yet. So my son got his license [in 2006], and they allowed him to have his grandfather, my father's, call sign again." During his son's flight, Garriott was able to respond to a contact from his father's call sign, now used by his son.

Garriott recalled another amusing communications-related incident during the STS-9 flight. Given the historic nature of the first flight of an international astronaut, a video link was set up between the shuttle, President Ronald Reagan, and West German chancellor Helmut Kohl. The link was to include Commander (and moonwalker) Young, German astronaut Merbold, and the U.S. payload specialist, Lichtenberg. In order to save the time of the three crew members to be featured, the two mission specialists and the pilot were asked to help set up the link and then turn it over to the others. In response to being treated as second-class astronauts, the trio made a silent protest—when the link was established for testing, the video showed Garriott, Parker, and Shaw in a classic "see no evil, hear no evil, speak no evil" pose, with one each covering his eyes, ears, and mouth. "What we were told was that was a hilarious, humorous scene in the control room and that it got into *Time* magazine."

While Garriott's launch experiences were fairly similar between the Saturn IB and the Space Shuttle, his landing experiences were not. Initially, he explained, the return to Earth begins similarly on both vehicles. A slight slowing causes the vehicle to dip into the atmosphere, creating an unforgettable glow as the air around the spacecraft heats up. "On the shuttle I happened to be sitting right beside the side window on the mid-deck, so you see all of the blue and yellow flames coming by. Then you get a little lower, it will turn to orange and reddish flames as the temperatures drop down a little bit."

At that point, the differences between the two reentries begin to be obvious.

As you really get lower, on the command module you have to pop drogue chutes to orient your spacecraft. When you get down to about ten thousand feet, I believe it is, the main chutes come out, and then you finally get down to a big splash in the water. That's quite different than coming down in a glider. You've been used to seeing the rate at which the Earth moves by [from space]. Once you get down to a lower altitude, the Earth starts moving by faster and faster, because you're lower, you're closer to it. So it's interesting to see those two comparisons.

The landing of STS-9 was more than a little unusual. According to Shaw, "We had another lesson on the landing of STS-9, and the lesson was, never let them change the software in the flight control system without having adequate opportunity to train with it."

The shuttle's flight control software interprets what actions it needs to take based on the inputs the commander makes on the stick, and how it responds depends on what point it is at in the mission. The settings that Young and Shaw used during training were replaced for the actual flight, which changed how the vehicle handled.

"I don't remember if we knew about that or didn't know about that," Shaw said,

> *but certainly when John started to de-rotate the vehicle, it responded differently than he had trained on. So here we are. John's flying the vehicle. I'm giving him all the altitude and airspeed calls and everything, and you feel this nice main gear gently settling onto the lake bed. . . . There were only two of us on the flight deck, as I recall, because we still had the ejection seats in* Columbia *at that time. They hadn't been taken out, so there was no room for another seat. So . . . the other four were on the mid-deck, and you hear this, "Yay!" and clapping when the main gear touched the ground very gently. Then John gets the thing de-rotated and we're down to about 150 knots or so when the nose gear hits the ground, and it goes "smash!" So it changes from this "Yay! Yay!" to "Jesus Christ! What was that?" That was just really funny, and I got all of that on tape, because I had a tape recorder going. And poor John, he was embarrassed because of this, the way the nose gear hit down, but it wasn't his fault. They had changed this thing without him being able to practice using the new flight control system. So that was a good lesson.*

The crew learned the next day that *Columbia* had also suffered yet another problem during the reentry and landing—a fire in the vehicle's auxiliary power units. "We had one APU shut down, and then when we shut the other two APUs down, normally after landing, it turns out one of them was also on fire," Shaw explained.

> *The reason the first one shut down was it was on fire . . . and it automatically shut itself down. . . . The next one didn't shut down until we actually shut it down. But there were two of them that were burning. . . . So we had a fire outside the APUs that when we shut them down and shut the ammonia off to them, the fires went out. So we had some damage back there, but the fires stopped.*

But we didn't know anything about that till the next day. I got this call and John says, "Hey, did you know that the APUs were burning?" "No, I sure didn't."

The flight also had two of the general-purpose computers (GPCs) fail. "That was an interesting thing, too," Shaw recalled.

John and I were in the de-orbit prep, . . . and we were reconfiguring the GPCs and the flight control systems and the RCS [reaction control system] jets and stuff. . . . About the time that we were reconfiguring the computers, we had a couple of thruster firings, and the big jets in the front fired and it's like these big cannons—boom! boom!—and it shocks the vehicle. You know, you really can feel it if you're touching the structure. So we had one of these firings and we got the big "X pole fail" on the CRT, meaning the computer had failed. This is the first computer failure we had on the program. Our eyes got about that big. So I get out the emergency procedure checklist and . . . we started going through the steps and everything. And in just a couple of minutes we had another one fail the same way. . . . So now we were really interested in what was going on. We ended up waiving off our de-orbit at that time. . . . The ground decided, no, we're going to wait and try to figure out what's going on with these computers.

When Young's shift ended, with nothing more he could do to address the problem and wanting to be rested for the landing, he went down to the mid-deck to take a nap, leaving Shaw "babysitting" on the flight deck.

Well, during that time frame, all of a sudden there starts this noise, bang, bang, bang, bang, bang, bang, bang, bang. The next thing, one of our three IMUs [inertial measurement units] . . . failed and we couldn't recover it. It turned out its gimbal was failing and it was beating itself to death against the gimbal stop, and that was the banging noise. After a few hours, John comes back upstairs and says, "You know, I really appreciate you guys making all that banging noise when I'm trying to sleep down there." I said, "Jeez, John, I've got some bad news. Man, we lost an IMU." And John's eyes get this big again, because we've had two GPC failures and now an IMU failure. Anyhow, we got through all that and we entered and landed, and when the nose gear slapped down, one of the GPCs that had recovered failed again. One of them didn't recover and we flew down with one less computer, but that computer failed again, and that's why I reconfigured flight control systems, as I remember now, because of that computer failure.

Despite the drama of the landing, once *Columbia* was successfully and safely on the ground, Shaw and the agency were pleased with the results of the shuttle's first dedicated scientific flight. "We learned a lot from that flight, a tremendous amount. Seventy-seven different investigations, as I recall, on that mission. It was a tremendous success."

STS-51B
Crew: Commander Bob Overmyer, Pilot Fred Gregory, Mission Specialists Don Lind, Norm Thagard, and Bill Thornton, Payload Specialists Lodewijk van den Berg and Taylor Wang
Orbiter: *Challenger*
Launched: 29 April 1985
Landed: 6 May 1985
Mission: Second Spacelab flight

The next Spacelab flight, 51B, came in April 1985. And there may have been no one more excited about it than astronaut Don Lind, who had been selected in NASA's fifth group of astronauts in 1966. "I set a record. No one has waited for a spaceflight longer than I have," Lind said. "I hope nobody ever has to do that. But with the six and a half years I spent in training for two flights that didn't fly, and then the delays in getting the shuttle program going, and with the [*Apollo 1*] fire, there were long delays, and so it was nineteen years before I got to fly."

For Fred Gregory, the pilot of the mission and one of the astronauts selected in the class of '78, the wait wasn't nearly so long, but it was nonetheless a huge honor to finally be selected for a crew. "There was shifting of launch times, because we were on Spacelab III and I know we launched before Spacelab II and after Spacelab I. I think they had some payloads that they wanted to deploy quickly, and so the laboratory missions were kind of put in a kind of a second category for priority. But, you know, the date wasn't important at the time."

Gregory found it interesting that the overwhelming majority of his training was for a very small part of the mission. Over three-fourths of his training focused on the eight and a half minutes of launch and hour of entry, out of a week-long mission. While the proportions may have seemed odd, it was necessary to make sure the flight deck crew was ready to perfectly execute the tasks required of it no matter what happened.

It's like a ballet, you know, without music: individual but coordinated activities that resulted in the successful accomplishments of each of these phases, regardless of the type of failure or series of failures that this training team would impose on you. So that's what we trained for. There were two thousand or so switches and gauges and circuit breakers, any number of which we would involve ourselves with during these two phases, ascent and entry. So the intent was for us to learn this so well, understand the system so well, that we could brush through a failure scenario and safe the orbiter in the ascent such that we could get on orbit and then have time to discuss what the real problem was and then allow you to correct it.

Eventually, it was time to move from the simulations to the real thing. As with STS-9, two shifts worked around the clock. "While each shift worked, the other shift slept," Gregory said.

We had enclosed bunks on the mid-deck of the orbiter, and that's where the off shift would sleep, so we never saw them really. . . . Each shift had its own area of interest, so there was really no competition between the two of them. . . . There were really about four hours a day there was an interaction between the two. It would be the two-hour postsleep transition, when one is just waking up. A shift is waking up, and they are picking up the ball, so to speak, from the other shift. The other shift then prepares to go to bed.

As pilot, Gregory's responsibilities in orbit mainly focused on the support systems that kept the orbiter functioning. He was technically in charge of one shift, but since the scientists were largely capable of carrying out their duties on their own, his support of them centered around maintaining the vehicle. "During that mission there were very few problems, and those that we had were very minor. So not only did I monitor the shuttle, but I also had a great deal of time to learn about the Earth, and then spent a lot of time looking into deep space. We were in a high-inclination orbit, 57½ degrees, and so it gave us an excellent view of a great part of the Earth."

The crew was joined by some rather unusual travel mates, recalled Don Lind:

We had the first laboratory animals in space, and Bill Thornton had to worry about them on one shift, and Norm would worry about them on the other shift. . . . We had two cute little squirrel monkeys and twenty-four less-than-cute laboratory rats.

27. The Spacelab in the cargo bay of the Space Shuttle. Courtesy NASA.

The squirrel monkeys adapted very quickly. They had been on centrifuges. They had been on vibration tables. So they knew what the roar and the feeling of space was going to be like. Squirrel monkeys have a very long tail, and if they get excited, they wrap the tail around themselves and hang on to the tip of the tail. If they get really excited, they chew on the end of their own tail. By the time we . . . activated the laboratory, which was about three hours after liftoff, they were now adjusted. They had, during liftoff, apparently chewed off about a quarter of an inch of the end of their tails, but they were adjusted and just having a ball. I kept saying, "Let's let one of them out." "No, no, can't do that. We'd never catch him again."

While the monkeys adapted quickly, the laboratory rats were somewhat slower. "The laboratory rats were not quite as savvy as the monkeys," Lind said. "They had also been on vibration tables and acoustical chambers and that sort of thing, but they hadn't learned that this was going to last awhile, and when we got into the laboratory, they were hanging on to the edge of the cage and looking very apprehensive. After about the second day, they finally found out if they'd let go of the screen, they wouldn't fall, and they probably enjoyed the rest of the mission. But they were slower in adapting. No big problem."

Gregory noted that he and Overmyer had the "privilege" of helping with the monkeys at one point, doing cleanup work capturing debris from the food and waste escape in the holding facility.

These rhesus monkeys that we had were extremely spoiled. I think that the environment that they had come from before they came on the orbiter was a place where they received a lot of attention from the caregivers there. Norm and I would look back into the Spacelab, and we would see Bill Thornton attempting to get these monkeys to do things, like touch the little trigger that would release the food pellets. And I could tell . . . watching them that they expected Bill to do that for them. So we looked back there one time, and we could see that kind of the roles were reversed, that Bill was actually doing antics on the outside of the cage and the monkeys were watching. We almost joked sometimes that they started laughing, and they went back and ate. It was an interesting dynamic to watch Bill Thornton wrestle with or react with the monkeys. It was quite an act back there.

Lind recalled that the substantial amount of automation involved in the Spacelab experiments provided him with a unique opportunity to do an experiment of his own. When he realized that a good bit of his duties would consist of regularly checking on automated equipment, he and a partner proposed an experiment to look at the aurora from space. Prior to his mission, the aurora had been photographed from space by slow-scan photometers, which give a blurred picture. On Skylab, Owen Garriott had taken a limited number of photographs of the aurora on the horizon.

"There were a few pictures, but not many," Lind explained.

So Tom and I started thinking about how we could do this. The first thing we wanted were high-time-resolution pictures of the aurora made with a TV camera. So we started looking around. What TV system could we get that would be sensitive enough in such a low light level? It turned out that the TV camera that was already on the Space Shuttle was as good as any TV camera we could have bought in the world. But we had to take off the color wheel and photograph in black and white instead of color. So we asked and got one of the TVs modi-

fied. Because [we would be doing that only] in black and white, you want to take some still photographs in color to document what color the auroral light is, since that will identify what particles are emitting the light. So we started to look around for an appropriate camera and camera lens. It turned out that the camera that we already had on board and the lens we already had on board were, again, as good as we could have gotten anywhere. NASA *only had to buy three rolls of film, special, sensitive color film. So this experiment cost* NASA *thirty-six dollars, and it's the cheapest experiment that has ever gone into space. It was very satisfying to us that we made some discoveries on thirty-six dollars. We claimed that we could do more science per dollar per pound than anybody else in the space program.*

The experiment proved to be not only cheap, but also effective, revealing a new component of aurora formation.

Despite the busy shift schedule and the additional experiment Lind had proposed to fill the extra time, he eventually found some time to appreciate the view from the orbiter's flight deck windows.

It was like a Cinerama presentation. Both my wife and I are amateur oil painters. . . . I thought, "Could I ever paint that?" The answer is absolutely not. Grumbacher [art materials company] doesn't make a blue that's deep enough for the great ocean trenches. You look out tangentially through the Earth's horizon, and you see—I was quite surprised—many different layers of intense blue colors, about twenty to twenty-two different layers of cobalt and cerulean and ultramarine and other shades, and then the blackest, blackest space you can imagine. When you go over the archipelagoes and the atolls and islands in the Pacific and down in the Bermudas, you see the water coming up toward the shore from the deep trenches, and it appears as hundreds of shades of blue and blue green up to a little white line, which is the surf, and another little brown line, which is the beach. Nobody will ever paint that. It's magnificent.

Lind was so overwhelmed by the beauty of the view that it brought tears to his eyes, which he discovered was also a very different experience in orbit.

In space, tears don't trickle down your cheeks; that's caused by gravity. In space the tears stay in the eye socket and get deeper and deeper, and after a minute or two I was looking through a half inch of salt water. I thought, "Ooh, this is like a guppy trying to see out of the top of the aquarium." Simultaneously with this, there's this sense of incredible beauty. But then I had a spiritual feeling, because several scriptures popped into my mind. You know, the nineteenth Psalm, "The heavens declare the glory of God." One of the unique Mormon scriptures is, "If you've seen the corner of heaven, you've seen God moving in his majesty and power." In the book of Romans, it says that the righteous will be coinheritors with Jesus Christ of all this. I thought, you know, "This must be the way the Lord looks down at the Earth." Because from space, you can't see any garbage along the highway. You can't hear any family fights. You see just the beauty of how the Lord created this Earth, and that was a very spiritual, moving experience, along with the aesthetics, the physical beauty. I'll always remember that special feeling, besides the technical satisfaction of a successful mission.

STS-51F

Crew: Commander Gordon Fullerton, Pilot Roy Bridges, Mission Specialists Story Musgrave, Tony England, and Karl Henize, Payload Specialists Loren Acton and John-David Bartoe

Orbiter: *Challenger*

Launched: 29 July 1985

Landed: 6 August 1985

Mission: Third flight of the Spacelab laboratory module

The third Spacelab flight, 51F, was commanded by Gordon Fullerton, marking his return to space after serving as pilot of the final demonstration flight, STS-4. "It was a great mission," Fullerton commented.

It really was. Some of the missions were just going up and punching out a satellite, and then they had three days with nothing to do and came back. We had a payload bay absolutely stuffed with telescopes and instruments. We had the instrument-pointing system that had never been flown. We had the idea of letting a satellite go and then flying this precise orbit around it and then going back and getting it. So, all kinds of new things, which took a lot of work to write the checklists for, write the flight plan, and so we spent a year and a half doing that.

After all the preparation, the mission had a rather inauspicious beginning. All three Space Shuttle main engines lit properly on schedule seconds before the scheduled launch, but then one failed before the solid rocket motors were ignited, leading to a launchpad abort.

"Karl Henize was pounding on his leg, really mad because he didn't get to go," Fullerton recalled.

I turned around to Karl and said, "We don't want to go, Karl. There's something wrong out there, you know." We were worried then: is this going to . . . mess everything up? It did to some extent, but the ground worked overtime, because everything was sequenced by time because it's an astronomy thing. Whether we're on the dark side or the light side, all that had to be rewritten. And it all worked out great. We even made up for the fuel we'd had to dump because of the engine failure on the way up [on our second attempt] and eked out an extra day on it. We were scheduled for seven and made it eight.

Once on orbit, the crew followed the established Spacelab pattern of working twenty-four hours a day, in two, twelve-hour shifts. "I anchored my schedule to overlap transitions, so if something came up on one shift, I could learn about it and carry it over to the next shift, hopefully," Fullerton recalled.

But I also had to stagger things so I got on the right shift for entry, so I was in some kind of reasonable shape at the end of the mission. At the beginning, too, we had the red team sleeping right up till launch time so that once we got on orbit, the red team was the first one up, and they'd go for it for twelve hours. So it was all that kind of thing, juggling around so that the right people that had to be alert for launch and entry were. We got into that circadian cycle prior to launch. So the last week they didn't see the other team; I only saw part of one and part of the other myself.

Compared to his first mission on a two-person demonstration flight, the presence of a larger, seven-member crew this time around had downsides as well as benefits.

The pressure is higher when you're commander—the pressure of making sure that not only you, but somebody else, doesn't throw the wrong switch. With Jack and I, it was just the two of us. He only had to worry about me, and I him. We could double-check each other. With seven people, there are many opportunities for somebody to blow it, not to say instant disaster, but to use too much fuel or to overheat some system or not have the right ones on and blow the chance to get this data. . . . That's a lot of other people throwing switches, too.

In addition to the responsibility of supervising the crew's actions, Fullerton found many other pressures involved in the role of commander.

During the entry, there was the pressure, [of] it's your fault if this doesn't come out right. When you're in the [pilot's] right seat, it's not all your fault. The commander bears culpability even if you make a mistake. I'm dwelling on this pressure thing because that really is a strong part of the challenge. I mean, you're really tired after spaceflight. I think you're tired mostly because you elevate yourself to this mental high level of awareness that you're maintaining. Even when you're trying to sleep, you're worried about this and that. So it's not like you're just lollygagging around and having a good time. You're always thinking about what's next and mostly clock watching. Flying in orbit is watching a clock. Everything's keyed to time, and so you're worried about missing something, being late.

The mission featured another historic footnote. In the mid-1980s the "Cola Wars" between competing soda brands Coca-Cola and Pepsi-Cola were in full swing, and on this mission the competition moved into space. Each company attempted to find a way to dispense its beverages in microgravity so that astronauts could drink them on the shuttle. Gravity plays an important role in the proper mix of carbonated sodas, and in microgravity the carbonation separates. The goal was to create a device that would ensure the beverages were properly mixed as they were consumed. Ultimately, however, the unique factors of the near-weightless environment limited the success of both attempts.

Astronaut Don Peterson, who had already left the astronaut corps at the time, recalled an incident during the 51F flight that demonstrated just how many people it took to make the shuttle successful, including a large number of people who never receive any recognition. "I remember, this . . . was one of the flights that Story Musgrave was on, because he and I were friends and I was kind of watching. During launch, they got an indication from instrumentation that one of the engines on the orbiter was overheating; . . . it was overpressure or overheat, something, and they shut it down," Peterson remembered.

They were far enough along in launch that they could still get to orbit. Well, then they got the same indication on the second engine. Now, if you shut down a second engine, you're into an abort, and that's a pretty messy operation. There was a young woman [who] looked at that. She was the booster control in the

Mission Control Center and this is happening in real time. You've [got] to realize, this sucker's up there burning away and you've got people, human beings, on board and all that. She looked at that and said, "I don't believe we've got two engine failures on the same flight. That's highly improbable. I believe we've got an instrumentation failure. Don't shut the engine down." Now, in hindsight, that was a wonderful decision, but had she been wrong, the back end of the vehicle would have blown out and killed everybody on board and lost a shuttle.

While the woman's decision saved the mission, she received little recognition for her split-second call, outside of NASA insiders closest to the situation.

Of course, the flight crew and all us people thanked her profusely, and she was recognized, but I don't think she ever really got any public recognition for that at all. But, I mean, that's a life-or-death decision under tremendous pressure with . . . human beings and a two-and-a-half-billion-dollar vehicle. And you can't get under much more pressure than that, and she called it right. What I did to her was terrible. I called her up later. I waited about a month and called her up and said, "I'm Robert Smith, and I'm a reporter with Life *magazine. I understand that you're the woman that saved Story Musgrave's life." And there was this long, long silence. And finally she said, "Who the hell is this?" Story has a reputation as a lady's man. So she kind of got a kick out of that, I think.*

STS-61A
Crew: Commander Hank Hartsfield, Pilot Steven Nagel, Mission Specialists Bonnie Dunbar, James Buchli, and Guion Bluford, Payload Specialists Reinhard Furrer, Ernst Messerschmid, and Wubbo Ockels
Orbiter: *Challenger*
Launched: 30 October 1985
Landed: 6 November 1985
Mission: German Spacelab laboratory flight

The fourth Spacelab flight, 61A, was yet another mission that introduced new elements to the shuttle program by building on its predecessors, particularly the first Spacelab flight. STS-9 had been distinguished from previous shuttle missions in part for carrying the first European crew member, West German astronaut Ulf Merbold. If STS-9 had been a further step toward international spaceflight, 61A was a leap—a mission purchased by and dedicated to the work of a foreign country, West Germany.

Commander Hank Hartsfield recalled explaining to a West German reporter the circumstances of Germany paying to use the German-built Spacelab module. "He said he wants to know how much Germany has to pay the United States to use their Spacelab, because Spacelab was built in Germany. It was built in Bremen. They were very sensitive about it. I think Germany had paid eighty million dollars for that flight. But this reporter was taking a very nationalistic look at it: 'We built it and now we have to pay to use it.'" Hartsfield explained that what the reporter failed to understand was that, while West Germany had built the modules, it did not own them. The first Spacelab module West Germany built, LM1, was given to the United States in exchange for flying Payload Specialist Ulf Merbold and experiments, and the second, LM2, was purchased by NASA.

According to Hartsfield, working with the Germans in planning the mission was an interesting process, filled with "delicate" negotiations. While he explained that he would stop short of describing NASA's West German partners as "demanding," they definitely had expectations for what they wanted out of the mission. "They pushed hard to get what they wanted out of the contract that they had signed with the U.S., and they took an approach where they would hang up on words, on what a word meant, in the agreement."

For example, one of the biggest controversies Hartsfield remembered is what language would be used on the mission.

They wanted to use the German language and talk to the ground crews in Germany and speak German. I opposed that for safety reasons. We can't have things going on in which my part of the payload crew can't understand what they're getting ready to do. It was clearly up front, the operational language will be English. We fought that one hard. We finally cut a deal that in special cases, where there was real urgencies, that we could have another language used, but before any action is taken, it has to be translated into English so that the commander or my other shift operator lead and the payload crew can understand it. . . . There were several times we did use the language during the flight, they asked for German, and it all worked real well. In fact, one case, I think it was national pride. Somebody wanted to talk to Wubbo [Ockels]—he was from the Netherlands—and they wanted to speak Dutch. Somebody insisted they had this urgent thing, and a friend of mine that spoke Dutch said, "You know what the guy wanted to do? He wanted to say, 'Hello, Wubbo, how's it going?'"

Hartsfield recalled a conversation with West German mission manager Hansulrich Steimle, whom he described as "a very interesting fellow" who "liked to philosophize." During preparation for the mission, the two became friends and would discuss cultural differences between the two countries. "He says, 'You know, in the United States, when a new policy comes down, the Americans, they look at this and say, 'Okay, here's what we've got to do,' and they salute and go do it. He says, 'In Germany, when a new policy comes down, we study it very carefully, and decide how we can continue to do what we're doing under this new policy.' I thought, 'Boy, is that ever true.' I know some people in the United States that do that, too, you know. I worked with some of them."

The multinational nature of the crew popped up in various and interesting ways, from the larger issues that led to the "delicate negotiations" to more trivial and potentially amusing incidents, Hartsfield recalled.

We were training in Building 5 [at Johnson Space Center], and once we went into quarantine, . . . we had to use the back door, and they issued keys to us. The keys had two letters and a number on it that identified that particular key. . . . Well, when we issued the keys, Ernst [Messerschmid, one of the German astronauts] came to me, and he was pulling my chain, because he's a wonderful personality, he said, "Is there any significance to the fact that I got a key with SS *on it?" ["*SS*" was the abbreviation for the Schutzstaffel, an official Nazi paramilitary organization during World War II.] So we got a big laugh out of it. He did. "No," I said, "it's the luck of the draw."*

The mission was the first to be partly directed from outside the United States. Germany had built its own control center in Oberphaffenhofen, southwest of Munich. Yet another of the interesting cultural differences came to light during a visit by the crew to the control center during mission preparations. "They had an integrated sim, the first one they'd run, and the whole payload crew was over there working in the simulator and doing this sim to get their controllers up to speed," Hartsfield recalled.

Well, they were also filming a documentary on how Germany was preparing for this thing, and so there were camera crews walking around in their control center taking pictures, and it got to be lunchtime. Things are different in Germany, you know, about drinking. To have wine and beer at lunch is a common thing. Well, in the basement of the control center, like in a lot of German businesses, they've got machines like Coke machines, but it's a beer machine. The flight controllers had all gone down and got a beer, and here is this crew, all of them got a beer, sitting on the console, and eating lunch. I called Hansulrich Steimle, and I said, "Hansulrich, I know how things are here in Germany, but you're filming for posterity here." I said, "If this film goes outside of Germany, some people may not understand your flight controllers drinking on duty." Then I became very unpopular, because some of them knew that I had said something, because he made them put their beer away.

The cultural differences were particularly an issue during mission preparations for Bonnie Dunbar, the only female member of the crew. As the two NASA mission specialists, she and Guy Bluford were sent over early to begin training in West Germany.

When I showed up there was a lot of discussion about both of us, but, with respect to me, they were very concerned that I had been assigned to the flight, because their medical experiment wasn't intended to include female blood. They thought that would ruin their statistics. . . . I was actually told in front of my face—and I have to first of all qualify that I've become very good friends with all these people; but any time you're at the point of the pathfinder, there's going to be things happening—I was told that maybe NASA had done this intentionally to offend the Germans by assigning a woman.

As they started to work through that, it came to light that some of the experiments on the mission, such as the vestibular sled experiment, did not fit her, and the Germans began saying they were going to need for her to be replaced. This was Dunbar's first flight assignment, and she began to worry that she would lose her seat. "At that time Dr. Joe Kerwin was head of Space and Life Sciences. . . . He actually wrote a memo for the record and to DFVLR [Deutsche Forschungs Versuchsanstat fur Luft und Raumfahrt, the German Aerospace Research Establishment] stating that all equipment should be designed to this percentile spread. So they stood by [me]."

However, Dunbar also had many positive experiences working with the West Germans in preparation for the mission and enjoyed their excitement about it. She began working with the Germans during preparations for the first Spacelab flight, and she said it was fun to see the atmosphere there during their entry into working with human-rated vehicles.

When I went over there, it was a very exciting time for them. The Bundestag had become involved in this, which is their legislative government. At that time the seat of government in [West] Germany was in Bonn, and we're training in Bonn/Cologne. So it wasn't unusual for our two German astronauts, Ernst Messerschmid and Reinhard Furrer, to have lunch with what we'd call a senator. There was just a lot of interest. The mission manager at that time told me that they hoped to use this mission not only to advance their science and human spaceflight but to inspire a generation of young people in Germany that really hadn't had inspiration since the war, and they lost the war. So this had very much a political flavor to it, not just a scientific flavor to it.

According to Dunbar, the German astronauts were great to work with, and it was very interesting being around them in West Germany, where they had a respected, heroic status.

Actually, it was fun to talk to them, because in Germany they very much occupied an Original [Mercury] Seven status, and so it was always interesting to hear from their perspective and see what they were doing, how the program was being received. It's changing now, but there's still a lot of dichotomy, even in Germany, about the value of human spaceflight, even though they're a major partner in ESA [the European Space Agency] and in the Space Station, because there would be one or two ministers that would vocally try to kill the program, and so we were interested in working with them and helping them.

Hartsfield recalled several highlights of 61A once the preparations were complete and the flight was finally underway. "I would say it was probably the most diverse mission ever flown," Hartsfield asserted.

We had a black, a woman, two Germans, a Dutchman, and a marine. I mean, how diverse can you get, you know? And there was some funny things happened. We launched on October 30, and of course, October 31 was Halloween. So I took a back off one of the ascent checklists we weren't going to use anymore, and I drew a face on it, cut out eyeholes, got some string, and I made myself a mask. I took one of the stowage bags and went trick-or-treating back in the lab. Of course, they don't do Halloween in Germany, or Europe, so they didn't know what I was up to. I decided not to pull any tricks on them, but I didn't get much in my bag. But somebody took a picture. One of the guys took a picture of me with that mask on, holding that bag, and somehow that picture got released back in the U.S. About a month after the flight, I got a letter from NASA headquarters. Actually, the letter had come from a congressman who had a complaint from one of his constituents about her tax money being spent to buy toys for astronauts. She was very upset. So it was sent to me to answer, and I had to explain, hey, nothing was done, and it was made in flight from material we didn't need anymore. It was just fun. I never heard any more, so I think maybe that satisfied her. She had the notion that we had bought this mask and bag and stuff just to do Halloween.

Another amusing anecdote Hartsfield recalled involved Bluford.

Steve [Nagel] and I were up on the flight deck. All of a sudden, we heard this bang, bang, bang! It sounded like somebody was tearing up the mid-deck. We peeked our heads down in the hole on the side where the bunks were. About that time we saw the bottom bunk come open, and the top one. Bonnie is sticking her head out looking up and Ernst is looking down, and all this banging is going [on] in this little bunk. So they slide Guy's door open to his bunk, and he kind of looks around, "Oh," and he pulls it back shut and goes back to sleep. . . . Apparently, he had awakened and didn't know where he was. He had a little claustrophobia or something, and he was completely disoriented, you know. But when he finally saw where he was, he said, "Okay," and he went back to sleep.

Hartsfield also recalled the crew flying through a mysterious cloud of particles during orbit. "Steve was up on the flight deck with Buckley, and they yelled back at me, 'Hey, Hank, get up on the flight deck.' And I got up there and looked out the window, and there was these little light things, bouncing off the windscreen. At first I couldn't tell how fast they were going. Zoom! Zoom! They looked like they were going real fast. And I said, well, it can't be that fast, and couldn't be massive, because they aren't breaking the window or anything. We were down close to Antarctica, at the southernmost point of our orbit."

For the duration of the mission, the crew had no idea what was causing the phenomenon. In fact, according to Hartsfield, answers didn't come until after the landing, when engineers looked at the orbiter's front windows. "When the solids [separated during launch], it would coat the windscreen. When those little things hit it, it was cleaning the windscreen. Those spots just took the grease right off of it. When we got back, we found out it was water. We had done a water dump on the previous rev. And we didn't realize it, but all of that turned into ice particles. And then we flew through it. It was weird. We were looking at the window, zoom zoom zoom. It would hit the windows and bounce off, and you're wondering what the hell it was."

10. Secret Missions

From its payload-carrying capacity to the wings that provided its substantial cross-range, the Space Shuttle was heavily shaped by the role the U.S. Department of Defense played in its origins. After Congress essentially pitted NASA's Skylab program and the air force's Manned Orbiting Laboratory against each other for funding in the late 1960s, NASA decided to try to avoid such problems with its next vehicle by soliciting DoD involvement from the beginning. In several ways, the shuttle's design and capabilities were influenced by uses the military had in mind for the vehicle.

Until early 1985, however, the military played only a limited role in the shuttle's use. Beginning as early as STS-4, there had been flights with classified military components, but there had yet to be a dedicated military classified flight. That would change with 51C, in January 1985.

STS-51C
Crew: Commander T. K. Mattingly, Pilot Loren Shriver, Mission Specialists Ellison Onizuka and James Buchli, Payload Specialist Gary Payton
Orbiter: *Challenger*
Launched: 24 January 1985
Landed: 27 January 1985
Mission: Launch of classified military intelligence satellite

T. K. Mattingly had dealt with classified elements previously, as commander of STS-4. After he returned from that mission, Deke Slayton asked him if he would be interested in staying in the astronaut corps and commanding the first fully DoD-dedicated classified mission. Slayton told Mattingly that the mission should require only six months of training, which would be a very short turnaround compared to many flights. Mattingly recalled, "With all the training and all of the years we put into the program, the idea of turning around and going right away was very appealing, to get my money back for all that time . . . and so I said, 'Yeah, I'll do that.'"

The pilot of 51C, Loren Shriver, said the rest of the crew was chosen to bridge the two worlds involved in the mission. Rounding out the crew were air force colonel Ellison Onizuka, marine corps colonel James Buchli, and air force major Gary Payton. "We knew that STS-10 [as it was originally called] was going to be DoD," Shriver noted, "and when the crew was formed, it was all military guys that formed the crew. I think NASA believed that it didn't have to do that, but I think it also believed that things would probably go a lot smoother if they did. So they named an all-active-duty military crew."

Although the promised quick turnaround had been the drawing card for the mission for Mattingly, problems with a solid-fuel engine used to deploy a payload on one of the shuttle's first operational missions caused a delay for the mission, since the plan was for 51C to also use a solid-fuel booster to deploy its classified payload. The flight was grounded for more than a year.

During the delay, Shriver learned that being assigned to a crew could have a downside. Traditionally, being named to a crew had been one of the best things that could happen to astronauts—they knew they were going to fly, they knew what their mission was going to be, and they had some idea of roughly when they would fly. With STS-10, which was renamed 51C during the delay, the crew discovered that sometimes being part of a crew could actually keep you from flying. "I thought, 'Well, maybe I never will fly.'" Shriver said. "It was the kind of situation where once you were identified as the crew for that mission, then especially this one being a DoD mission, you were kind of linked to it, as long as there was some thought that it was going to happen. And it never did completely go away. It just went kind of inactive for a while and then came back as 51C."

The classified payload for the mission was reportedly the Magnum satellite, a National Security Agency satellite used to monitor military transmissions from the Soviet Union and China. While the mission was officially classified, according to news reports at the time, information about the payload and its purpose is available in congressional testimony and technical journals.

Everything about the mission was classified, not just the payload. This included all details about training, astronaut travel, and even the launch date. "I couldn't go home and tell my wife what we were doing, anything about the mission," Mattingly said. "Everybody else's mission, everybody in the world knew exactly what was going on; NASA's system is so wide open. They could tell their wives about it, their family knew, everybody else in the world knew what was on those missions. We couldn't talk about anything. We couldn't say what we were doing, what we had, what we were not doing, anything that would imply the launch date, the launch time, the trajectory, the inclination, the altitude, anything about what we were doing in training. All that was classified. Couldn't talk about anything."

People ask questions all the time, Mattingly said, and they ask even more questions when they know that they can't know the answer.

Then they just get even more adamant that you should tell them and try to dream up of more tricky ways to get you to say something—the media, of course, being number one in that game. . . . Everybody had an opinion as to what it was, and you'd just say, "Cannot confirm or deny," and that's all that was necessary. . . . It was humorous, I guess, to listen to people out there trying to guess as to what it might be. You'd say, "Okay. Well, just let that churn around out there. I'm going to go do my training and not worry about it." And eventually you don't think much about it. But it does require you, then, when you do go meet the press or you do go do public presentations, that you have to think a little bit harder about what you can say and not say.

As a result of the delay, Mattingly and crew spent a substantial amount of time preparing for the mission, in an unusual experience that blended the very different cultures of NASA and the military. "The interesting thing about the classified mission is JSC and the whole NASA team has worked so hard at building a system that insists on clear, timely communication," Mattingly said. "The business is so complex that we can't afford to have secrets. We can't afford to have people that might not know about something, even if it's not an anomaly. For something that's different, something unusual, we try to make sure that it's known in case it means something to somebody in this integrated vehicle."

But for a classified military mission the number of people who could know and talk about the mission details was limited. "I had some apprehensions about could we keep the exchange of information timely and clear in this small community when everybody around us is telling anything they want, and we're kind of keeping these secrets," Mattingly explained. "Security was the challenge of the mission. How do you plan for it? How do you protect things? We went around putting cipher locks on all the training facilities, but then you had to give the code to a thousand people so you could go to work."

Shriver also recalled concerns that the requirement of classification of elements of the mission would interfere with the open communication that was a vital part of safety and mission assurance. "In the NASA system, everything is completely open; . . . everybody is pretty well assured of having the information that they know they need. We were concerned that just the opposite was going to happen, that because of the classification surrounding the mission, people were going to start keeping secrets from each other and that there was a potential that some important product or piece of information might not get circulated as it should."

In reality, Shriver opined, NASA and the DoD managed to find ways to make compromises that protected the information that needed to be protected while allowing for sharing of information that needed to be shared.

The mission, yes, was classified. Certain descriptive details about what was going on were always classified, but within that classification shell, so to speak, the system was able to find a way to operate and operate very efficiently, I thought. There were still some hiccoughs here and there about how data got passed back and forth, and who could be around for training and who couldn't, and that sort of thing. But eventually all that got worked through fairly well, and I think the pathfinding we did on that mission helped some of the subsequent DoD-focused missions to be able to go a little bit smoother.

A special room for storing classified documents and having classified conversations was added to the Astronaut Office. The new ready room included a classified telephone line. "They said, 'If certain people need to get hold of you, they'll call this number,'" Mattingly recalled.

It's not listed and it's not in the telephone book or anything. It's an unlisted number, and this causes less attention. "You've got to keep this out of sight, don't let anybody know you've got it, and this is how we'll talk to you on very sensitive things." So we had a little desk in there, put it in a drawer, and closed it up. In the year we worked on that mission, we spent a lot of hours in this little room because it was the only place we could lay our stuff out; the phone rang once, and yes, they wanted to know if I'd like to buy MCI [telephone] service.

Even if, by and large, the mission stakeholders did a good job making sure that the secrecy didn't prevent needed information from being shared, there was at least one occasion when it went too far, Mattingly recalled. "My secretary came in one day, and she was getting used to the idea that there's a lot of people we deal with that she doesn't know. . . . She came in to me one day, and she says, 'You just got an urgent call.' 'Okay.' 'Joe (or somebody) says call immediately.' So, okay. 'Joe who?' She said, 'He wouldn't tell me. He said you'd know.' We went in our little classified room and said, 'Does anybody know a Joe?' We never did figure out who it was, and he never called back."

Preparing for a classified mission was quite the adjustment because of the differences between NASA and air force security systems and the many rules for how to deal with classified information. "In any bureaucracy we sometimes overdo things, but as much [as] we make fun of these folks, they convinced me that some of the precautions we were taking were, in fact, justified," Mattingly stated.

I was a bit skeptical, but they showed me some things that at least I bought into. Whenever we traveled, they wanted to keep secret when was the launch time, and they certainly wanted to keep secret what the payload mission was. And to keep the payload mission secret, that meant whenever we went somewhere they wanted us to not make an easy trail when we'd go somewhere. To keep the launch time classified, they wanted us to make all our training as much in the daytime as at night, so that someone observing us wouldn't be able to figure this out. They never convinced me that anyone cared, but they did convince me that if you watch these signatures you could figure it out, and it is secret because we said it was.

The extra work to preserve secrecy, in Mattingly's opinion, turned out to have some benefits, but also plenty of downsides to go along with them.

I didn't mind the idea of flying equal [time in] day and night, because that meant I got to fly more, because I wasn't about to split the time, we'll just double it. So that was a good deal. But then they had this idea they wanted us, whenever we went to a contractor that was associated with the payload or with the people we were working with, they didn't want us to get in our airplane and fly to that location. They wanted us to file [flight plans] to go to Denver and then refile in flight and divert to a new place so that somebody who was tracking our flight plans wouldn't know. And when we'd get there, we could check in using our own names at the motel, but, you know, just Tom, Dick, and Harry. So, just keep a low profile.

Even so, the mission preparations did, at times, demonstrate the difficulty of keeping secrets when too many people know the information.

We went out to Sunnyvale [California], and we were going to a series of classes out there, and this was supposed to be one of these where you don't tell anybody where you're going, don't tell your family where you're going to be, just go. But the secretary got a room for us. So we went, landed at one place, went over to another place, landed out there at Ames, had this junky old car that could hardly run. El [Onizuka] was driving, and Loren and Jim [Buchli] and I were crammed in this little tiny thing, and we're going down the road and looking for a motel. And we didn't stay in the motel we normally would stay at. They put us up and tell us to go to some other place and they had given us a name. So we went to this other place, and it was very inconvenient and quite a ways out of the way. And as we drive up the road, Buchli looks out the window and he says, "Stop here." So we pull over, and he says, "Now let's go over [the security procedures] one more time. We made extra stops to make sure that we wouldn't come here directly, and they can't trace our flight plan. And we didn't tell our families, we didn't tell anybody where we are. And we can't tell anybody who we're visiting." He says, "Look at that motel. What does that marquee say?" "Welcome STS-51C astronauts," and everybody's name is in it, and you walk in and your pictures are on the wall. Says, "How's that for security?"

Security lapses such as that proved frustrating for the crew, Mattingly recalled, when they themselves made great efforts to preserve secrecy only to watch the information get out anyway. "Those are dumb things, but they show that we went to extraordinary lengths trying to learn how to do some of these things. And the coup de grâce came when, after, you know, 'I'll cut my tongue off if I ever tell anybody what this payload is,' and some air force guy in the Pentagon decides to hold a briefing and tell them, before we launched, after we'd done all these crazy things. God knows how much money we spent on various security precautions and things."

Even with the things that were revealed, the public face of the classified mission was also unusual, Mattingly said, explaining that the public affairs people at Mission Control had an interesting time dealing with the media. "For the first time the MOCR's [Mission Operations Control Room] not going to be open for visitors, there's nothing to say, nothing to do, you know. 'They launched.' 'Yeah, we saw that.' 'Oh, they came back.' 'That's good.'"

On launch day, those tuning in heard the usual launch discussions at launch control, but the communications between ground control and the astronauts were not broadcast, as they had been on all previous flights.

Shriver said that the secrecy had personal ramifications for the astronauts, who wanted to share with their friends and family their excitement about the flight—in Shriver's case, his first.

The air force did not even want the launch date released. They didn't want the crew member names released. We weren't going to be able to invite guests for the launch in the beginning. This is your lifetime dream and ambition. You're finally an astronaut, and you're going to go fly the Space Shuttle, and you can't invite anybody to come watch. It was an interesting process. We finally got them talked into letting us invite [people]; I think each one of us could invite thirty people, and then maybe some other car-pass guests who could drive out on the causeway. But trying to decide who, among all of your relatives and your wife's relatives, are going to be among the thirty who get to come see the launch, well, it's a career-limiting kind of decision if you make the wrong decision.

All in all, Mattingly said, he considered the mission to be a success and was proud to be a part of it. He said the real contributors were the people who prepared the payloads but that it was an honor to have the opportunity to deliver them.

I still can't talk about what the missions were, but I can tell you that I've been around a lot of classified stuff, and most of it is overclassified by lots. I think at best it's classified to protect the owners, you know, it's self-protection. What those programs did are spectacular, they are worth classifying, and when the books are written and somebody finally comes out and tells that chapter, everybody is going to be proud. Now, all the things we did for security didn't add one bit, not one bit. But the missions were worth doing, really were. The work was done by others, but just to know that you had a chance to participate in something that was that magnificent is really kind of interesting.

STS-51J

Crew: Commander Bo Bobko, Pilot Ron Grabe, Mission Specialists David Hilmers and Bob Stewart, Payload Specialist William Pailes

Orbiter: *Atlantis*

Launched: 3 October 1985

Landed: 7 October 1985

Mission: Deployment of two military communications satellites, first flight of *Atlantis*

Bo Bobko was assigned to command the next classified military mission, 51J, which flew in October 1985. Like 51C, its crew also had a strong military presence: Pilot Ron Grabe of the U.S. Air Force, David Hilmers, U.S. Marines, Bob Stewart, U.S. Army, and Payload Specialist William Pailes, Air Force.

Despite the mission's classified nature, Commander Bo Bobko described the mission as "pretty vanilla." "I mean, we went on time and we landed according to the schedule," Bobko said. "The fact that it was classified was a pain, but you lived with that. Somebody might be doing an experiment, and he could be working on the experiment out in the open as long as it was away from NASA, but having the experiment associated with that shuttle flight was the classified part of it. So I couldn't call a person, because as the commander, if I called them, it would give an indication that that experiment was on that shuttle flight."

The 51J mission was the first flight of *Atlantis*. Bobko, who had also served on the maiden voyage of *Challenger*, said *Atlantis* flew well on its first flight.

NASA has subsequently released the information that the payload for the mission included a pair of Defense Satellite Communications System satellites, part of a constellation of satellites placed in geosynchronous orbit to provide high-volume, secure voice and data communications. The system was a next-generation upgrade from a network the military originally began launching in 1966.

11. People and Payloads

In 1983 a new classification of astronauts had emerged, joining pilots and mission specialists: payload specialists. Until this point, being an astronaut was a full-time job, a career choice that people committed to for years. They were given broad training, which prepared them to carry out any variety of mission they might be assigned. Payload specialists, on the other hand, had other careers; flying in space, for them, was not their job, but a job duty. Payload specialists were just that—specialists from organizations, including universities, companies, and nations, responsible for a shuttle payload, who accompanied that payload during flight and helped with its operation.

On 28 November 1983 the STS-9 mission, also known as Spacelab I, had launched into space the first payload specialists: Byron Lichtenberg, a biomedical engineer from the Massachusetts Institute of Technology, and Ulf Merbold, a West German physicist representing the European Space Agency. Lichtenberg was the first American who was not a career astronaut to fly in an orbiting U.S. spacecraft, carrying out an experiment in space that he helped design and that he would help analyze and interpret as a member of a research team. Before that, scientists had instructed and trained astronauts on how to do their experiments and astronauts did the work for them.

STS-41D
Crew: Commander Hank Hartsfield, Pilot Michael Coats, Mission Specialists Judy Resnik, Steven Hawley, and Mike Mullane, Payload Specialist Charles Walker
Orbiter: *Discovery*
Launched: 30 August 1984
Landed: 5 September 1984
Mission: Deployment of three satellites

The year after Lichtenberg and Merbold's flight, the first commercial payload specialist, Charlie Walker, flew as a member of the 41D crew. Commanded by Hank Hartsfield, 41D deployed three satellites and tested the use of a giant solar wing. Walker was assigned to the flight to run the Continuous Flow Electrophoresis System (CFES), an apparatus from the McDonnell Douglas Corporation, for which Walker was a test engineer. The system used electrophoresis, which is the process of separating and purifying biological cells and proteins.

"What they were producing with that was erythropoietin," Hartsfield recalled. "It's a hormone that stimulates the production of red blood cells. Ortho Pharmaceuticals was the primary contractor with MacDac to build this thing. And CFES was kind of a test version of it. . . . The idea was, which was a good one, say you were going to have planned surgery, they could inject that hormone into you prior to the surgery, some time period, I don't know how long it would take, but your body would produce more red blood cells, and then you wouldn't need transfusions. So it was a good idea."

Prior to being assigned a flight of his own, Walker was responsible for training NASA astronaut crews in the operation of the CFES payload on the STS-4, STS-6, STS-7, and STS-8 shuttle flights during 1982 and 1983. He flew with the CFES equipment as a crew member on 41D, 51D, and 61B.

The initial agreement between McDonnell Douglas and NASA called for six proof-of-concept flights of the CFES device, which was originally to be flown aboard Spacelab. But according to Walker, slips in the launch of Spacelab caused McDonnell Douglas and NASA to renegotiate for six flights on board the Space Shuttle mid-deck.

The agreement was that NASA would provide the launch and McDonnell Douglas would provide the equipment and testing processes. "NASA, in the body of the Marshall Space Flight Center Materials Lab folks, had the opportunity, at no expense to them other than the preparation of samples and then the collection and the analysis of it later in their own laboratories upon return home, to use a device produced at the expense of the private sector for private-sector research, but, again, allowing NASA the right to use it for up to a third of the time in orbit in exchange for the opportunity to have it there in orbit aboard shuttle," Walker explained.

After the first flight of the CFES device, on STS-4, Walker said McDonnell Douglas felt that it had demonstrated enough success in the proof of concept test to ask to fly its own astronaut to run its device. "From our standpoint, we had proven that we could predict adequately for production processing what we needed to know," Walker said.

We briefed on that, and we advised the Space Shuttle program management what we wanted to do for the next flight; got that approved through appropriate processes, and at the same time—I can remember, I was in a meeting in which Jim Rose [McDonnell Douglas manager] and I briefed Glynn Lunney [shuttle program manager] in Glynn's office in Building 1 [at JSC], and Jim told me, going down, he said, "I just want to tell you, as we walk into this meeting if I get an indication from Glynn that he's happy with the results, too, from the NASA side, I'm going to ask for a payload specialist opportunity." He said, "Are you okay with that?" And I said, "You know I'm okay with that."

That was exactly what happened, Walker recalled. Lunney indicated that NASA was pleased with the results, so Rose pressed ahead.

Jim Rose said, "We want to ask for the opportunity to negotiate for one more thing." And basically it went something like, "You know, Glynn, the astronauts that we're training, Hank's a great individual, obviously a great test pilot, a good engineer, but Hank doesn't know this electrophoresis stuff, and the other astronauts, mission specialists that we're going to train, they're going to be able just to spend a little bit of their time working with our device. They've got lots of other things to do. That's the mandate for the mission specialist. We really would learn the most we possibly can and more than we can do with a mission specialist if we get the opportunity to have a payload specialist devoted specifically to the electrophoresis device and its research and development activities during a flight."

As I remember it, Glynn chewed on his cigar a little bit—that's when you [could] smoke in the office—chewed on his cigar a little bit and said something like, "Well, we've been wanting to move into this payload specialist thing, so if you've got somebody that is qualified, can meet all the astronaut selection criteria, put in the application. Let's do it." . . . I think Glynn said something like, "Do you have somebody in mind?" Jim turned to me and looked at me, and Jim said something like, "You're looking at him." Glynn said, "You, huh?" And I said, "Yes, it's me. I'll be the man."

Walker was added to the crew for 41D in late May 1983. Within a few days of being added, Walker, who was residing in St. Louis at the time, went down to Johnson Space Center to officially meet the crew, some of whom he knew from the training he had provided for previous missions, and to begin working out a training schedule. The training syllabus was based loosely on the training developed for science payload specialists on Spacelab missions and was to be a foundation for future commercial payload specialists. The training syllabus was the result of ongoing negotiations between Walker, his employers, and various stakeholders at NASA. Walker enjoyed the training and felt it was important. Unlike other astronauts, however, for whom mission training was their job, training took Walker away from his day-job duties in St. Louis, so while his employers wanted him trained, they also wanted him available there when it wasn't necessary for him to be in Houston. NASA also wanted to make sure he received all the training needed to fly safely but didn't want to invest unnecessary time and resources that were needed for career astronauts.

Walker explained that the payload specialist training was a shorter, condensed version of the career astronaut training. Initially, he said, that condensed training still involved at least an overview of a wide variety of systems. "Hank Hartsfield had me operating the remote manipulator system, the RMS, in the trainer, and that went on for a few weeks. I was training with the crew. I was working the RMS in the simulator, and I knew the system, I knew how to work it, even though I was not in the flight plan to deploy any of the satellites or to have to use the RMS, as might conditionally be the case. I don't think on [41D] we had any required use of the RMS, but we had as a contingent the operation of it."

Later, however, management came back and decided that Walker didn't need to be trained on equipment he wasn't going to use. There was no need, they argued, to invest valuable time and resources training payload specialists on things they wouldn't be doing. Walker said that regardless of the actual need for the additional training, he felt like it had substantial benefits for the mission. "My comments were that I think this is a good thing," he said. "Let the payload specialist do some of this, too. He or she is going to feel like more of a cohesive part of the crew. It's just a good psychological thing, even though you don't need their hands to especially do that."

The most important and most time-consuming part of preparing for a mission was for payload specialists to gel with the crew, Walker said. "In other words, the crew's getting to know me, and me getting to know my fellow crewmates for each flight, so that we knew, at least to a significant degree, each other's characteristics, and we could work together and feel good working together and flying together as a team."

Walker spent, on average, about two weeks each month training at JSC during the nine months leading up to the flight. He said Hartsfield wanted to integrate him with the crew as closely possible, and in addition to official training sessions, the commander included him in the occasional social event as well. "I was invited to more than one dinner or activity at Hank's house and some of the other homes of the astronauts—the crew as well as others—but it wasn't as close a relationship as was the case between the career astronauts and families down there that were obviously living and working at each other's elbows day in and day out."

Hartsfield said he felt so strongly that Walker should be an integral part of the crew that he requested Walker's name be on the patch circle with the rest of the crew. "[Payload specialists'] flight assignments changed a lot," Hartsfield said. "Some of the flights had as many as three different people assigned at one time or another, and they had to keep changing their patches. So to save money, they put a ribbon at the bottom with the payload specialist on it, so they wouldn't have to change the whole patch. But when Charlie flew, I had sold George Abbey on this. 'He's part of the crew, you know. Put his name on the patch with the rest of us.'"

Admittedly prejudiced, Hartsfield described the 41D crew as one of the best crews ever put together. "As the commander, I just sort of had to stay out of their way," he said. "I was reminded of that two-billed hat, you know, that says, 'I'm their leader. Which way did they go?'"

The addition of payload specialists brought a new dynamic to spaceflight and the astronaut corps. Walker's crewmate Mike Mullane said several astronauts, including himself, had viewed payload specialists as outsiders and as competitors for flight assignments. "There was some friction there, I think, that we felt like, 'Hey, why isn't a mission specialist doing this experiment,'" said Mullane. "But I'm mature enough now, and particularly after you get your one mission under the belt, you become a little more tolerant of the outsiders."

28. The crew of STS-41D. Courtesy NASA.

Walker said there were only a few individuals in the agency from whom he got the impression that he and the other payload specialists were considered outsiders. "I was there as a working passenger. I wasn't a full-fledged crew member, and I knew that going in, and I took no real exception to that," Walker said. "Occasionally there were circumstances in which it was made clear to me that 'You're not one of us. You're along for the ride, and you've got a job to do.' But it was only a few individuals, some in the Astronaut Office, others outside the Astronaut Office, from whom I got that impression."

One clear distinction between the career astronauts and the payload specialists was the locations of their offices. At Johnson Space Center, the payload specialist office was in Building 39, and the Astronaut Office was in Building 4. According to Walker, "It was made clear to us from the beginning that they didn't expect to see us over on the fourth floor in Building 4 except for scheduled meetings. We were just outsiders who would become crew members for a short period of time and would train mostly on our own, but when there was necessary crew combined training, certainly we would be there."

Walker said he was fine with that arrangement and was just grateful for the opportunity, both professionally and personally. "There was no belligerence, really, expressed openly, and no offense on my part taken," Walker said.

I really saw my role and my place in this, this was a great adventure, and more than an adventure, it's a great challenge, both to people as well as to technical systems. I think I know my limitations, and I know that I'm not nearly as qualified to make critical and rapid decisions in some of these flight environment circumstances, as the men and women that have been selected by the agency through grueling processes, to do just exactly that. . . . I was getting a great opportunity, I felt, both for the company that was my employer, for the commercial as well as the prospective societal benefits from the work that we were proposing to do through and with the Space Shuttle. And certainly, certainly a tremendous personal opportunity for me, and I was just happy to be there.

The crew went out to launch on 26 June 1984, and everything was going smoothly until the clock reached four seconds. "The engines had already started to come up, and then they just shut down," Hartsfield said. "We looked at the countdown clock on the onboard computer display, and it was stopped at four seconds. We were really checking to see if there was anything out of the ordinary. We were going to make sure that things were still okay. There was a good moment of tension there, and Hawley broke the tension. As soon as we looked at everything and everything was okay, Steve said, 'Gee, I kind of thought we'd be a little higher at MECO [main engine cutoff].'"

As the Launch Control Center was trying to figure out the problem, a hydrogen fire was noticed on the pad. "The trouble with a hydrogen fire is you can't see it," said Hartsfield. "Hydrogen and oxygen burn clear, and you can just see some heat ripples when you're looking through it, but you can't see the flame. I think one of the sensors picked it up, a UV sensor, which can see it."

When the fire was discovered, there was talk about having the crew leave the shuttle via the slide wire Emergency Egress System. No one had ever ridden the slide wire, Hartsfield said, and flight controllers were afraid to tell the crew to do it. "That bothered a lot of us in that they were concerned enough about the fire that they really wanted us to do an emergency egress from the pad area, but since the slide wire had never been ridden by a real live person—they'd thrown sandbags in it and let it go down—they were afraid to use it, which was a bad situation, really."

Hawley remembered the crew members talking about whether or not they should get out of the orbiter and use the Emergency Egress System baskets. "I remember thinking, well, fire's not too bad because then you're sitting inside this structure that's designed to take several thousand degrees during reentry. It's well insulated. Then I got to thinking, on the other hand, it's attached to millions of gallons of rocket fuel, so maybe that's not so good. But eventually they came and got us."

As a result of the pad abort, NASA revamped and tested the procedures for the Emergency Egress System and implemented new training for aborting and recycling launches.

Steve Hawley said Mullane was very concerned after the fire that the flight was going to be canceled. "I really didn't think they would do that, but I remember him being very concerned about that, probably more concerned than [about] the incident itself. He was concerned about the effect it would have on his flight assignment."

While the flight did get delayed for two months and its payloads were changed as a result, the mission did not get canceled entirely. Mullane said he found scrubs personally and emotionally draining. "There is nothing that is more exhausting than being pulled out of that cockpit and knowing you have to do it tomorrow," he said. "It is the most emotionally draining experience I ever had in my life of actually flying on the shuttle. I will admit that it is terrifying to launch. Once you get up there, it's relaxing, but launch, it's terrifying. And people assume that it gets easier. I tell people, no, it doesn't. I was terrified my first launch. I was terrified my second launch. I was terrified my third launch. And if I flew a hundred, I'd be terrified on a hundred."

Mullane said that before every launch he felt like he faced the possibility of death, as if he were preparing to die. "I know it's ridiculous to think you can predict your death," Mullane said.

You could get in an auto accident driving out to get in the T-38, and that's your death, and here you are thinking it's going to be on a shuttle. But I certainly prepared for death in ways, in a formal way. I served in Vietnam, and there was certainly a sense of you might not come back from that. And I said my goodbyes to my parents and to my wife and young kids when I did that, but this time it was different because it's such a discrete event. It's not like in combat where in some missions you go off and fly and never see any enemy antiaircraft fire or anything. But this one you knew that it was going to be a very dangerous thing. And as a result, twenty-four hours before launch, you go to that beach house and you say goodbye to your family, to the wife, at least. That is incredibly emotional and draining, because the wife knows that it could be the last time she's ever going to see you, and you know it's the last time you might ever see her.

Weeks before launch, crew members and their families choose a family escort to help families with launch details and to be with them during launch. The family escort stands next to the family on top of the Launch Control Center during launch. Part of that role is simply helping the family get to where they need to be, making sure everything goes smoothly. However, in addition, the family escorts serve as casualty assistance officers in the event something unexpected happens.

Mullane recalled that his wife commented to him, "'What I'm picking isn't a family escort; it's an escort into widowhood.' You have this buildup, this incredible emotional investment in these launches that just ticks with that clock. Picking the astronaut escort. The goodbye on the beach house, at the beach house, that lonely beach out there. And now to go and get into the cockpit. Like I said, I thought a lot about death. I mean, I felt this was the most dangerous thing I would ever do in my life was ride this shuttle."

Mullane opined that it was a mistake on NASA's part to build the shuttle without an escape system. "I don't know what the thought process was to think that we could build this rocket and not need an escape system, but it was the first high-performance vehicle I was ever going to fly on with no escape system," said the three-time mission specialist.

If something went wrong, you were dead. So that was the sense of death that kind of rode along with you as you're driving, preparing for this mission and driving out to the launchpad. You know it's the most dangerous thing that you've ever done in your life. And to get strapped in and be waiting for that launch, and man, I'll tell you, your heart is in your throat. I mean, after a launch abort, I swear, you could take a gun and point it right at somebody's forehead, and they're not even going to blink, because they don't have any adrenaline left in them; it's all been used up. To be strapped in out there and then to be told, "Oh, the weather's bad. We got a mechanical problem," and to be pulled out of the cockpit, and now it's all going to start over. Twenty-four hours you go back, you're exhausted, you go back, have a shower, meet your wife, say goodbye again, and then start the process all over the next day. And you do that two or three times in a row, and you're ready for the funny farm. It really is a very emotionally draining thing.

The 41D mission had a total of three scrubs—two in June and one more in August—and finally lifted off on 30 August 1984.

Once the crew was finally in orbit, Steve Hawley, for whom 41D was his first flight, said it took several days to adjust to microgravity, but the team didn't have several days before starting to work. "It's interesting because what we've always done . . . is plan the mission so that the most important things happen first," Hawley said.

That goes back to the days when we'd not flown the shuttle before and everybody was concerned that it was going to fall out of the sky, and so if you got up there, you needed to get rid of the satellite or whatever it was right away, so that when a problem happened, you'd have the mission accomplished. But the shuttle is very reliable, and so what you end up doing is doing the most important, most challenging, most difficult tasks when the crew is the least prepared to do it, because they're inefficient and they haven't adapted yet. . . . Back in those days, we were launching satellites five hours after we got on orbit, and we were still trying to figure out how to stay right side up.

The 41D mission deployed three satellites: two Payload Assist Modules and a SYNCOM for the navy. It was the first time three satellites were launched on one flight. The mission also performed a demonstration of the Office of Application and Space Technology solar wing, referred to as OAST-1. The 102-foot-tall, 13-foot-wide wing carried different types of solar cells. It demonstrated the use of large, lightweight solar arrays for future use in building large facilities in space, such as a space station.

As part of the demonstration, OAST-1 was extended to its full height several times, stretching out of a canister mounted on a truss in the payload bay. "When fully extended, it was 102 feet tall, and really spectacular to look at," Hartsfield said.

The array did not have very many actual solar cells; instead, it was primarily a test structure to see how well the truss would extend. It featured three linear rods with cross-rods and cables, such that the rods were in compression and the cables were in tension. The structure collapsed into a cylindrical canister for launch. NASA engineers had predicted how rigid the structure would be based on models, and the orbital experiment would give them the opportunity to validate those predictions and models.

"Surprisingly, once the thing is deployed, it's fairly rigid," Hartsfield said. "What was interesting was the array was an order of magnitude stiffer than the engineers had predicted, which was a big surprise to them. In fact, by the time we got ready to fire the second set of firings, which was supposed to increase the motion, the thing had almost stopped completely, it was so stiff."

Walker's Continuous Flow Electrophoresis System worked as planned, but postflight analysis showed that the samples had some biological contamination. "In other words, a little bit of bacteria had gotten into some of the fluids during preparation before flight, and the bacteria had grown during flight and contaminated what we intended to have as . . . biologically pure, uncontaminated by extraneous bacteria," Walker said. "So the work that I had done had been, so to speak, technically productive. We learned new procedures. We validated the procedures. But the veracity of the biological sample itself for the medical testing that we were going to do postflight turned out to be a problem, turned out to be bad. So we were not a complete success in terms of our mission accomplishment because of that."

One of the more memorable episodes on the flight was the infamous "peecicle." During flight, the crew had a problem with an icicle forming around the nozzle where they dumped wastewater, primarily urine and condensation from the orbiter's humidity control. There was a lot of concern about the icicle because when the orbiter started reentry the frozen water was in just about the right place to break off and hit the Orbital Maneuvering System, Hartsfield said. "If you hit the OMS pod and broke those tiles, that's a real high-heat area right on the front of that pod, you could burn through. And if you burned through, that's where the propellant is for the OMS engines, and that's not a good thing to have happen."

Hawley recalled that the ground called up and had the crew test the wastewater dump. "I think we didn't know anything was unusual initially," Hawley said.

I think maybe the ground called us and told us to terminate the supply water dump because they had seen some temperature funnies. So we did, and then sometime later, I guess they got curious enough to use the cameras on the robot arm to see what was there. So we set the arm up, and yes, you could see this icicle there. For whatever reason subsequent to that, they decided that we ought to try a waste dump and watch it with the camera on the arm, and the icicle was still there. I remember, as we were doing it, watching the second icicle form. So we ended up stopping that dump, and now here we are with this icicle.

The ground crew started working to find possible solutions, one of which was to turn that side of the orbiter toward the sun and let the icicle melt. "After about three days we were convinced that the ice was not going to sublime off the orbiter," Hartsfield said. "It reduced in size somewhat, but it was still there. I had people ask me, 'Gee whiz, you got it right in the sun, why didn't it melt?' I said, 'The same reason snow and ice don't melt on a mountain. It's in direct sunlight, but it doesn't absorb much heat. It reflects most of it.' That was the same thing as this icicle. It wasn't going anywhere."

The next option was to send astronauts on a spacewalk to break off the ice. Hawley and Mullane had trained as contingency EVA crew members and were selected for the EVA, if there was to be one. "I remember Mike was thrilled," Hawley said,

because he was going to get to do a spacewalk, and I'm sitting there going, "This is not a good idea. I don't know how in the world we're going to get to it." I mean, it's down on the side of the orbiter aft of the hatch, and there's no trans-

lation path down there. I guess they were talking about taking the CFES unit apart, using some of the poles that the CFES was constructed with to maybe grab one of us by the boots and hang him over the side and have him knock it off. That all sounded like a bad plan to me.

It was decided not to try an EVA but to use the robotic arm to knock the ice off instead. "I remember thinking, 'Yeah, it's a good plan,'" Hawley recalled, "and Mike was thinking, 'Oh no, I'm not going to get to do an EVA.'"

While mission controllers were trying to resolve how to get rid of the icicle, the crew faced another, more immediate issue inside the orbiter. To avoid making the icicle larger, the decision was made that the crew would not be able to dump the waste tank again. While there was still some room in the tank, calculations revealed that the condensation that would be collected during the rest of the mission would fill that volume. "What that meant practically to us," Hawley explained, "was that we couldn't use the toilet anymore, because there was no room in the waste tank for the liquid waste."

The crew members collected their waste in plastic bags and stored the waste-filled bags on board. Walker said some of the bags were left over from the Apollo program. "I'm kind of an amateur historian," Walker said, "so I felt a little bad at peeing in these historic bags, but we had to do what we had to do."

In retrospect, Hartsfield said, the incident is funny, but it wasn't funny at the time. "The problem was that in zero g, Newton's third law is very apparent to you. If you just try to use a bag, when the urine hit the bottom of the bag, it turned around and came right back out, because there's no gravity to keep it there. Didn't take long to figure that wasn't going to work." The astronauts stuffed the bags with dirty underwear, socks, towels, and washcloths to absorb the urine.

Hartsfield decreed that the only female on the crew would continue to use the shuttle bathroom. "Judy, as you can imagine, had a hard time with the bag, so we had a little room in there. I said, 'I don't care what the ground says, you use the bathroom. The rest of us will do the bag trick.'"

The situation was messy, with all of the bags being stored in the waste storage tank under the floor. Hartsfield said there was at least one instance where a crew member was stuffing a urine bag into the tank and the bag ruptured.

Twice, Hartsfield said, Flight Director Randy Stone asked management if the crew could convert a water tank to a waste tank. It would have been an easy conversion, Hartsfield said, but at the time there was great concern about turning the orbiters around quickly for the next flights and the response was that using the water tank as a waste tank would add a week to the process of getting the orbiter ready to fly again.

"I sometimes think I made a mistake," said Hartsfield.

I probably should have called for a private med conference and told Flight, "Hey, we've got a real problem up here." . . . I talked to the guy that headed that room up when we got back, and he apologized. They later found that it wouldn't have impacted the flow at all. I said, "Joe, you just don't know what we're going through up there." "Well, you should have told somebody." "I don't want to put that on the loop." I mean, in fact, Gerry Griffin, the center director, when we got back, he expressed his thanks for not putting that on the open. The media would have had a ball with that.

[Initially] we were hoping to stay another day on orbit, because we had enough fuel to do it, but this was not a very good situation. By the time day six came, we were ready to come home.

For the second time, on board the shuttle was an IMAX camera. Hawley recalled that the IMAX camera pulled film so fast that in zero g it would torque the user like a gyroscope. The camera had a belt drive with a belt guard, but for this flight, it was decided the belt guard wasn't needed. "I don't know if we were trying to save weight or what, but we decided we didn't need this belt guard." Hawley said. "I'm up there doing something, and all of a sudden I hear this blood-curdling scream. I go floating upstairs to see what had happened, and Judy had gotten her hair caught in this belt for this IMAX camera, and there was film and hair all over the orbiter. It jammed the camera and the camera blew the circuit breaker that it was plugged into."

Mullane said he, Resnik, and Hawley were filming the SYNCOM launch when the incident happened. Mullane and Mike Coats cut Resnik's hair to free her from the camera, and Coats then spent hours picking hair out of the camera gears in order to get the camera working again. The crew dealt with the problem on their own, without reporting it to the ground, concerned that if the public found out, the incident would provide fodder for those critical of NASA's flying female astronauts.

29. Judy Resnik with several cameras floating around her, including the IMAX camera in which her hair got tangled. Courtesy NASA.

Throughout the mission, each crew member went about his or her assigned tasks with very few coordinated crew activities, Hartsfield said. As a result, he made sure that they ate dinner together every night. "You get a quick breakfast snack, and the first thing you know, you're off on your daily do list," Hartsfield said, describing a typical day during the mission. "You eat lunch, normally, on the run where you've got a lull in your activities. But I had decreed that the evening meal we were all going to eat together. I want one time for the crew to just get together and just chat and have a little fun and say, 'Okay, where are we? What have we got to do tomorrow?' and talk about things."

One night during dinner as a crew, a rather strange thing happened. "We'd prepared our meals, and we were all floating around, holding down on the mid-deck, and all of sudden we heard this knocking noise, like somebody wanted in," Hartsfield recalled.

It sounded like knocking. Holy crap, what is that? And then we had a traffic jam trying to get through the bulkhead to get up to the flight deck, because it was coming from up that way somewhere. So we got up there, and we were on the night side of the Earth, it was pitch black out there. Steve flipped on the payload bay lights. You know those housings take like five minutes. And we said,

"God, what is that!?" We could hear it, whatever it was, was on the starboard side. Steve was the first one to see it. He looked at the gimbal angles on the Ku-Band antenna, and it was banging back and forth. It was something where it was oscillating back and forth. He hit the power switch and turned it off, and that did it. And we went, whew. And we told the ground later what had happened, and it never did it anymore, whatever it did. Apparently it got into a range where it kept trying to swap or do something. I never did find out exactly what caused it, but it sure got our attention. Some alien wants in.

The mission lasted six days, and then it was time to come home. Walker said the reentry and landing on 41D was an emotional experience, drawing to a close what he thought at the time was a once-in-a-lifetime experience.

At that point I didn't know I was going to have any further flights. I thought that was probably it, and it was such an extraordinary experience, and now it was, for sure, over with. I came to sense a real defining moment, a physically and emotionally defining moment. This experience, this great thing called spaceflight, . . . probably above everything else, it's based upon velocity. It's putting people and machinery at high speed at the right velocity, the right altitude, the right speed, around the Earth till you keep going, and you're working in this high-velocity environment that we call orbital flight. When you want to come home, you just take out some of that velocity with some rocket energy again, and use the Earth's atmosphere to slow you down the rest of the way until you come gliding in and lose the last part of the velocity by applying brakes on the runway until you come to a stop.

So I noted in my own mind two definitive points here that really, without debate, start and end this great experience. One is the high-energy event that we call launch, straight up when the rockets start; to the landing and wheels stop on the runway horizontal, and the brakes have taken hold, and the energy is gone, and the spaceship literally rolls to a stop.

With the end of the mission came the completion of the first flight of a commercial payload specialist and an opportunity to evaluate how the idea actually worked in reality. Hawley commented on how well the crew worked together and how well the crew got along with Walker. "It's more important who you fly with than what your mission is, and we really had a good time," Hawley said. "We all got along well. I thought we all had respect for each other's capabilities, and it was just a good mix. . . . Charlie was a good guy. He fit in very well. We enjoyed having him as part of the crew."

The flight marked the beginning of the process, over time, of the softening of hard lines between the career astronauts and the payload specialists. A major milestone, Walker said, was the decision to move the payload specialists into office space with the career astronauts. "Even while I was training for 61B, I had office space. It was, oh, by the way, catch it as you can, but you got office space over on the fourth floor, Building 4. You need a place to sit and work when you're in town, come on over. Finally they moved the PS Office out of Building 39 over to Building 4 in that time period just before *Challenger* was lost."

The integration of noncareer astronauts became even more complicated as NASA implemented plans to fly even more types of people on the Space Shuttle—academic and industrial payload specialists, U.S. politicians, international payload specialists, and the first "Teacher in Space" and "Journalist in Space."

"I think the clearest example as an indicator of how things transformed was to follow the Teacher in Space activity, because originally the Teacher in Space was to be a spaceflight participant/payload specialist, and I witnessed a lot of slicing and dicing of just what do you call Christa McAuliffe," Walker recalled. "Is she an astronaut? Well, most people at the time at JSC and certainly in the Astronaut Office were, 'No, she is not an astronaut. We were selected by NASA to be astronauts. We're the astronauts. She's a payload specialist.'"

STS-51D

Crew: Commander Bo Bobko, Pilot Don Williams, Mission Specialists Rhea Seddon, Jeffrey Hoffman, and David Griggs, Payload Specialists Charlie Walker and Senator Jake Garn

Orbiter: *Discovery*

Launched: 12 April 1985

Landed: 19 April 1985

Mission: Deployment of two satellites

Despite feeling like his first flight would be a once-in-a-lifetime experience, eight months later Charlie Walker was back in space, this time on 51D. The mission deployed two satellites and carried into space several science experiments and yet another payload specialist. This time, in addition to Walker, on the crew was Jake Garn, a U.S. senator from Utah and the first elected official to fly aboard the Space Shuttle.

Garn was added to the crew about two to three months before launch, recalled Commander Bo Bobko. "George Abbey said to me one day, he said, 'What sort of training program would you have if you had a new passenger that was only going to have eight or twelve weeks?'" Bobko recalled. "I said, 'Why are you asking me that question?' He said, 'Because you've got a new passenger, and you've only got—,' I don't know, ten or twelve weeks to flight."

Walker said Garn had been lobbying for some time with the NASA administrator to get a chance to make a flight on the Space Shuttle. Garn was chairman of a NASA oversight body within the Appropriations Committee of the U.S. Senate. "Just part of his job; he needed to do it," Walker said.

Of course, you look at Senator Garn's history, and at that point he had some ten thousand hours logged in I don't know how many different kinds of aircraft, having learned to fly as a naval aviator, and had gone to the air force when the navy tried to take his ticket away from him and wouldn't let him fly again. . . . Jake was very aviation oriented and certainly enamored with the agency's activities and just wanted to take the opportunity if one could be found. So his lobbying paid off, and he got the chance to fly. He was still in the Senate and would take the opportunity on weekends to come train down here; would take congressional recesses, and instead of going back to his home state, to Utah, he'd come down here to JSC. So he worked his training in and around Senate schedules.

Walker recalled hearing some negative talk around the Astronaut Office after Garn was added to the crew. "I do remember that there was at least hall talk around the Astronaut Office of, 'Oh, my gosh, now what's happening here to us? What have we got to put up with now?'" Walker said.

But Jake, from my experience, and here is an outsider talking about another outsider, but I think Jake accommodated himself extraordinarily well in the circumstances. . . . What I saw was a Jake Garn that literally opened himself up to, "Hey, I know my place. I'm just a participant. Just tell me what to do, and I'll be there when I need to be there, and I'll do what I'll [need to] do, and I'll shut up when I need to shut up," And he did, so I think he worked out ex-

traordinarily well, and quite frankly, I think the U.S. space program, NASA, *has benefited a lot from both his experience and his firsthand relation of* NASA *and the program back on Capitol Hill. As a firsthand participant in the program, he brought tremendous credibility back to Capitol Hill, and that's helped a lot. He's always been a friend of the agency and its programs.*

Bobko lauded Garn for knowing what it meant to be part of a crew. "I'd call him up and I'd say, 'Jake, we need you down here.' And he'd say, 'Yes, sir,' and he'd be down the next day for the sim," Bobko said.

Garn's only problem, added Bobko, was that he got very sick on orbit.

He was doing some of these medical experiments, and they find that one of the things that happens is that on orbit, if you get sick, your alimentary canal, your digestive system, seems to close down. So what they had were little microphones on a belt that Jake had strapped to him to see if they could detect the bowel sounds. So the story is—and I haven't heard it myself—they had me on the microphone saying to Jake, "Jake, you've got to get upstairs and let them see you on TV. *Otherwise, they'll think you died and I threw you overboard."*

According to Walker, he and Garn were the guinea pigs for quite a few of the experiments on the flight. One of the experiments was the first flight of a U.S. echocardiograph device. "Rhea [Seddon] was going to do echocardiography of the hearts of I think at least three of the crew members, and of course, Jake and I were the obvious subjects," said Walker. "We really didn't have much of a choice in whether we were going to be subjects or not. 'You're a payload specialist; you're going to be a subject.'"

While in the crew quarters prior to launch, Walker said, Garn was asking him the typical rookie questions about what it's going to be like and what to expect.

He says, "You've done this before. Tell me. Give me the real inside scoop. What's this going to feel like? What's it going to be like?" "It's going to be great, Jake. It's just going to be great. Just stay calm and enjoy it." . . . We got into orbit, and I can remember there was the usual over-the-intercom exuberant pronouncements, "Yee-ha, we're in space," yadda, yadda, yadda.

I can remember shaking hands, my right hand probably with Jake's left, gloves on, and "We're here," and then Jake and I both kind of look at each other, and we're both beginning to feel weightlessness.

The crew was originally assigned as 51E, but that mission was canceled and the payloads were remanifested as 51D. The mission deployed a communications satellite and SYNCOM IV-3 (also called LEASAT-3). But the spacecraft sequencer on the SYNCOM IV-3 failed to initiate after deployment. The mission was extended two days to make certain the sequencer start lever was in the proper position. Griggs and Hoffman performed a spacewalk to attach flyswatter-like devices to the remote manipulator system. Rhea Seddon then used the shuttle's remote manipulator system to engage the satellite lever, but the postdeployment sequence still did not start.

"Once it became clear that there was a problem, we got a little depressed," Walker said. "You train for these things to happen. You know they're really important. Here's hundreds of millions of dollars' worth of satellite out there. Your flight's not that inexpensive, of course, to send people into space. So a lot of effort has gone into getting this thing up there and to launching it and to turning it on and having it operate, in this case, for the United States Navy. And here it didn't happen, so you're like, 'Oh, my gosh.'"

The crew immediately began to think in terms of contingencies. Walker recalled a strong awareness of the nearness of the satellite. Despite the fact that it had failed, it was still there, floating not that far away. It was still reachable and could potentially be repaired or recovered.

Within a few days the ground came up with the suspected culprit—a mechanical switch, about the size of a finger, on the side of the satellite was supposed to have switched the timer on. The thought was that maybe that switch just needed to be flipped into the right position. If the shuttle could rendezvous with the satellite, all that would be needed would be some way to flip the switch.

The ground crew instructed the flight crew to fashion two tools that Walker referred to as the "flyswatter" and the "lacrosse stick." "The ground had faxed up to us some sketchy designs for these tools, and I think there were two tools that were made up. I can remember cutting up some plastic covers of some procedures books. We went around the cabin, all trying to find the piece parts, and the ground was helping us."

Working together, the ground and the crew in space began an *Apollo 13*–like effort to improvise, using available materials to fashion a solution to the problem, Walker said. "The in-flight maintenance folks on the ground were, of course, very aware of what tools were on board, and they looked down the long list of everything that was manifest and tried to come up with a scheme of what pieces could be taken from here, there, and anywhere else on board, put together, and to make up these tools for swatting the satellite."

The shuttle rendezvoused with the satellite, and Hoffman and Griggs exited the shuttle for the EVA.

These guys go outside, and they're oohing and aahing about the whole experience and doing great. . . . Rhea commands the remote manipulator system over to the side of the cargo bay. Literally with more duct tape and some cinching straps, they strap the flyswatter and the lacrosse stick on the end of the remote manipulator arm. Then they come back inside, and we make sure they're okay, and they secure the suits. Bo and Don finish rendezvousing with the satellite, and Rhea very carefully moves the two tools on the end of the RMS right up against the edge of the satellite.

Walker noted that none of these procedures had been rehearsed on the ground; it was all improvised using the various skills the crew had picked up during their training. "This was all done just with the skills that the crew had been trained with generically, the generic operation of the remote manipulator system, the generic EVA skills, and the generic piloting skills to rendezvous with another spacecraft," Walker said. "And yet we pulled it off; the crew pulled it off expertly, did everything, including throwing the switch."

Bobko said the crew had not done a rendezvous simulation or any rendezvous training in several months, and the books with rendezvous instructions weren't even on board. "So they sent us up this long teleprinted message, and I've got a picture of me at the teleprinter with just paper wound all around me floating there in orbit," Bobko said. "It turned out to be a rather different mission. But, luckily, in training for the missions that had been scheduled before, we had learned all the skills that were required to do this. If we had just trained for this mission, we probably wouldn't have ever trained to do a rendezvous or the other things that were required."

Unfortunately, flipping the switch didn't take care of the problem and there was nothing else the crew could do at that point. However, the ground was able to determine that the problem was with the electronics and the satellite would need to be fixed on a subsequent flight. (It ultimately was repaired on the 51I mission.) "We felt a little dismayed that the satellite failed on our watch and that we weren't able to fix it on the same flight, but we felt gratified that we took one big step to finding out what the problem was, that eventually did lead to its successful deployment," Walker said.

While the problems with the primary payload weren't discovered until they got into the mission, another payload—Walker's electrophoresis experiment—had encountered difficulties much earlier. About three days before launch, while the orbiter team was preparing *Discovery* for flight, Walker and several McDonnell Douglas folks were working with the CFES equipment when it started to leak. "My project folks were out there filling it full of fluid, sterilizing it with a liquid sterilant, and then loading on board the sample material and then the several tens of liters of carrier fluid," Walker said. "That electrophoresis device started leaking. Inside the orbiter, on the launchpad, it started leaking. Drip, drip, drip. Well, of course, that didn't go over very well with anybody, and our folks diligently worked to resolve that. Right down to like twenty-four hours before flight or so, that thing was leaking out on the pad."

Program managers began discussing whether the leak could be overcome so that the device could be loaded for operation. If not, the experiment could not be conducted during the mission. "The question became, 'Well, maybe we don't even fly Walker, if he doesn't have a reason to fly,'" Walker recalled. "So there was active discussion until about a day before flight—this is all happening within about a twenty-four-hour period up till T minus twenty-four or thereabouts—as to whether I would fly or not, because maybe my device wasn't going to be operational in flight and so I had nothing to do, so to speak. But it was resolved."

The leak was fixed, the fluid was loaded, and the equipment—and Walker—were cleared for launch. In flight, the CFES worked well. And, in addition to running the CFES apparatus, during the mission Walker conducted the first protein crystal growth experiment in space, a major milestone in biotechnology research. "This was the first flight of the U.S. protein crystal growth apparatus," he said. "Actually, it was a small prototype that Dr. Charlie Bugg from the University of Alabama Birmingham and his then-associate, Larry DeLucas, had designed and had come to NASA, saying, 'We've got this great idea for the rational design of proteins, but we need to crystallize these and bring the crystals back from space. We think they'll crystallize much better in space, and we can do things up there we can't do on Earth, etc., etc., but we need to fly it on board a Space Shuttle flight to see if it will work.'"

The flight was also the first for the NASA Education Toys in Space activities, a study of the behavior of simple toys in a weightless environment. The project provided schoolchildren with a series of experiments they could do in their classrooms using a variety of toys that demonstrate the laws of physics. Astronauts conducted the experiments with the toys in orbit and videotaped their results. Students could then compare their results to what actually happened in space. The toys flown included gyroscopes, balls and jacks, yo-yos, paddle balls, Wheelos, and Hot Wheels cars and tracks. "I still to this day feel a little chagrined that I wasn't offered a toy or the opportunity," Walker said. "Everybody else had a toy, but not me. . . . Even Jake Garn had paper airplanes."

Walker may not have played with toys, but he played with liquids, conducting some fluid physics experiments with supplies on the orbiter.

Jeff Hoffman and I spent one hour preparing, at one point later in the mission, some drinking containers, one with strawberry drink and one with lemonade. . . . We would each squeeze out a sphere maybe about as big as a golf ball of liquid, floating in the cabin, and we actually played a little game in which we would put the spheres of liquid in free floating, oh, about a foot apart from each other, and Jeff and I would get on either side, and somebody would say, "Go." We'd start blowing at the spheres with our breath, just blowing on them, and we'd try to get them together and get them to merge, because it was really cool when they merged. One big sphere suddenly appears that's half red and half green, and then the internal fluid forces would start to mix them, and it's really interesting to watch.

As the astronauts were blowing, their breath would actually move their bodies around. At the same time, the balls of liquid would start going in different directions, and the two together would make it increasingly difficult to keep the liquid under control. "You've got to be quick," Walker said, "and usually there's got to be somebody with a towel standing by, because either a wall or a floor or a person is going to end up probably getting some juice all over them."

Walker said he felt more comfortable going in to his second flight than his first. "Not to say that I felt blasé or ho-hum about it, by no means," he said.

You just can't go out and sit on a rocket and go into space and feel ho-hum about it, even after umpteen flights. It just isn't going to happen. But a person can feel more comfortable. Some of the sharp edges, to put that term on it, of the unknown, of the tension, are just not there. I guess maybe a better way to put it, I would suggest, is now you really know when to be scared.

The second time around you're not focusing on the same things. You're now maybe a little less anticipatory of everything. You know [how] some things are going to be, so you can kind of sift those and put those aside in your mind and pay attention to other aspects. There were other things that I paid attention to, like I maybe was more observant of the Earth when I had a chance to look out the windows, more sensitive to the view.

Walker recalled Jeff Hoffman sharing with the crew his interest in astronomy, and in particular the crew trying to spot Halley's Comet. "There was one or more nights, . . . in which we turned off all the lights in the cabin and night-adapted our eyes. Everything was dark. . . . I can remember us trying to find the Halley's Comet and never feeling like we succeeded at doing that. But, it was still so far away and so dim that it really probably wasn't possible. But just looking at the sky along with an astronomer there was a great and tremendously interesting experience."

Landing was delayed by a day, giving the crew an extra day in space, which Walker said he spent mostly looking out the window observing Earth. "I just never got bored at looking at the ever-changing world below," Walker said. "You're traveling over it at five miles per second, so you're always seeing a new or different part of the world, and even [as] days go by and you orbit over the same part of the world, the weather would be different, the lighting angles would be different over that part of the world. Just watching the stars come up and set at the edge of the Earth through the atmosphere, watching thunderstorms."

During the landing at Kennedy, *Discovery* blew a tire, resulting in extensive brake damage that prompted the landing of future flights at Edwards Air Force Base until the implementation of nose-wheel steering. Walker said the landing at first was just like the landing on his first mission. "Things were again just as they'd been before and as was planned and programmed, so no big surprises until those final few seconds when you expect to be thrown up against your straps by the end of the braking on the runway and the stop. Well, in our case, we're rolling along about ready to stop, and then there's a BANG, and I can remember Rhea looking at me, and Jake saying, 'What's that?'"

Walker said one of the tires had locked up, skidded, and scuffed off a dozen layers of rubber and insulation and fiber until the tire pressure forced the tire to pop. "It ended up just a little bit off the center line of the runway because of that, but we were going very slow, so there was no risk of running off the runway at that speed because of the tire blow. But certainly we heard it on board, and there was a thump, thump, thump, and we stop. We were going, like, 'Well, what was that?' I don't know; in my own mind, I was thinking, 'Did we run over an alligator? What happened here?'"

STS-51G
Crew: Commander Dan Brandenstein, Pilot John Creighton, Mission Specialists Shannon Lucid, John Fabian, and Steven Nagel, Payload Specialists Patrick Baudry (France) and Prince Sultan Salman Al-Saud (Saudi Arabia)
Orbiter: *Discovery*
Launched: 17 June 1985
Landed: 24 June 1985
Mission: Deployment of three communications satellites, test of SPARTAN-1

Like so many missions before it, 51G succumbed to mission, crew, and payload shuffling. Commander Dan Brandenstein said shuffling like that was just how things were at this point in the shuttle program. There was a lot of scrambling around with missions for a variety of reasons, and the program was still relatively new, Brandenstein said.

That was early '85. We had only been flying four years. The vehicle hadn't matured as you see it today. So they were flying technical problems on a vehicle and they'd have to pull one off the pad. That affected shuffling and payloads didn't come along quite like they figured, and that affected shuffling. So it was sort of a variety of things. . . . Then we got canceled and picked up these four satellites. We had one for Mexico, one for the Arab Sat Consortium, one for AT&T, and then we had SPARTAN, which was run out of Goddard. It was one that we deployed and then came back and recovered two days later. So it was a lot of mission planning changed and we had a couple new crew people that we had to integrate into the crew and all that.

With three satellites to deploy into orbit, the 51G crew deployed one satellite a day for the first three days on orbit. "Shannon and I had the lead on those deployments and J. O. Creighton was flying the orbiter, so he was pointing it in the right directions and so forth," recalled Mission Specialist John Fabian. "Brandenstein was making sure that everybody was doing the right things. That's what a commander is supposed to do. And Sultan was taking pictures for his satellite. I mean, it was a fairly routine operation."

The SPARTAN proved to be a little more challenging. The SPARTAN spacecraft were a series of experiments carried up by the Space Shuttle. The program was based on the idea of a simple, low-cost platform that could be deployed from the Space Shuttle for a two- to three-day flight. The satellite would then be recovered and returned to Earth.

"It was a much simpler satellite," Fabian said, "from the crew's perspective, than the SPAS-01 [a German satellite that Fabian released and recaptured using the robotic arm on STS-7] because the SPAS-01, we could maneuver it. It had experiments on it that we could operate, had cameras on it that we could run. The SPARTAN, which was a navy satellite, we simply released it, let it go about its business, and then later went back and got it."

Shannon Lucid did the release, and two days later Fabian did the recapture. Deployment was routine, Fabian said. "At least it appeared to be," explained Fabian. "When we left it, it was in the proper attitude. It was an x-ray astronomy telescope, and while we were gone, it took images of a black hole, which is kind of cool stuff. That's kind of sexy."

But when the orbiter came back to retrieve it, Fabian said, the satellite was out of attitude. The grapple fixture was in the wrong position for the shuttle's arm to be able to easily grab it.

One idea was to fly an out-of-plane maneuver, flying the shuttle around the satellite, but the crew hadn't practiced anything like that. Fabian noted,

Dan's a very capable pilot, and I'm sure that he could have done that, but it turns out that perhaps an easier way would be to fly the satellite in much closer to the shuttle, get it essentially down almost into the cargo bay, and then reach over the top with the arm and grab it from the top, and that's what we elected to do.

Of course, we told the ground what was going on, that it was out of attitude, and they worried, but there wasn't much they could do—they couldn't put it in attitude—so they concurred with the plan, and that's what we executed.

Fabian said it felt good to benefit from all of the time spent in the simulators with the robotic arm. Training for contingency situations contributed greatly to the crew's knowledge of the arm's capabilities and to the successful retrieval, he said.

The seven-person crew for this mission, including two payload specialists and representing three different nationalities, had a unique set of challenges because of those factors, said Fabian, but in general the crew got along well and had a positive experience.

We were told not to tell any camel jokes when Sultan showed up, and the first thing he did when he walked through the door was to say, "I left my camel outside." So much for the public affairs part of the thing. These just were not issues. They really were not issues. Patrick flew a little bit of French food and didn't eat the same diet that we ate. Sultan did. Patrick flew some small bottles of wine that were never opened, but the press worried about whether or not they had been. Patrick flew as a Frenchman and enjoyed it, I think.

Fabian and Nagel were assigned to support Baudry and Sultan with any help they needed on their experiments. "Patrick was doing echocardiographs," Fabian said, "and he did those on himself, and he did them on Sultan, and I think he did them on one or two of the NASA crew members, and frankly, I've forgotten whether he did one on me or not. But he was using a French instrument with a French protocol, and it was the principal thing that he was doing in flight, was to do these French medical experiments."

Sultan's primary role was to observe, Fabian explained.

We were flying $130 million worth of satellites for the Arab League. But he also had some experiments, and he was tasked to take pictures, particularly over Saudi Arabia, which of course would be very valuable when he got home. People would be very interested in seeing that. But they didn't need a lot of support. They didn't need a lot of help. We had to worry a bit about making sure that they were fed and making sure that they knew how to use the toilet and making sure that they understood the safety precautions that were there and so forth. And, you know, probably more than half of what our role and responsibility was with regard to the two. Other than that, it was to make sure they had film when they needed it in the cameras and help them for setup if they needed some setup for video or something of that type and to participate in their experiments to the degree that it was deemed necessary.

Even at this point, the payload specialist classification was still very new and crew members were still figuring out exactly how to act toward each other, and that resulted in a change being made to the orbiter. "People weren't really sure how these folks were going to react," Fabian recalled. "We put a lock on the door of the side hatch. It was installed when we got into orbit so that the door could not be opened from the inside and commit hari-kari, kill the whole crew. That was not because of anybody we had on our flight but because of a concern about someone who had flown before."

Fabian expressed concern over how the agency handled safety during this era of the shuttle program. On this flight, for example, Fabian said the ARABSAT never passed a safety review. "It failed every one of its safety reviews," Fabian said.

The crew recommended that it not be flown, the flight controllers recommended that it not be flown, and the safety office recommended that it not be flown, but NASA management decided to fly it. This was an unhealthy environment within the agency. We were taking risks that we shouldn't have been taking. We were shoving people onto the crews late in the process so they were never fully integrated into the operation of the shuttle. And there was a mentality that we were simply filling another 747 with people and having it take off from Chicago to Los Angeles, and this is not that kind of vehicle. But that's the way it was being treated at that time.

It was very disappointing to a lot of people, a lot of people at the agency, to see management decide to fly this satellite. But if they hadn't flown the satellite, you see, political embarrassment, what are we going to do with the Saudi prince, what about the French astronaut, what's the French government going to have to say about us saying that we can't fly their satellite on the shuttle, what will be the impact downstream of other commercial ventures that we want to do with the shuttle? Well, of course, after Challenger, *the commercial all went away, and it was a dead-end street anyhow, but we didn't know it at the time.*

12. The Golden Age

How different it was in those early years of shuttle, when we were going to fly once a month at least. That was going to be routine, and we were going to revolutionize space and discover these amazing things, and we still will, but we were just naive, thinking it was going to happen the next year, and not the next decade or the next generation. So there was a lot of naiveté, and maybe it was just us or maybe it was just me, but that was the big change. It's a little sad that that had to happen, but that's just maturing the industry, I guess.

—*Astronaut Mike Lounge*

STS-51I

Crew: Commander Joe Engle, Pilot Dick Covey, Mission Specialists Ox van Hoften, Mike Lounge, and Bill Fisher

Orbiter: *Discovery*

Launched: 27 August 1985

Landed: 3 September 1985

Mission: Deployment of three satellites; retrieval, repair, and redeployment of SYNCOM IV-3

Like the 51A mission, which recovered two satellites that had previously failed to deploy, 51I included the repair of a malfunctioning satellite from a previous mission, 51D. The satellite, SYNCOM IV-3, had failed to activate properly after deployment. Mike Lounge recalled that he and fellow 51I mission specialist Ox van Hoften were together when they heard about the SYNCOM failure. The satellite was fine; the failure was a power switch on the computer. "We, essentially on the back of an envelope, said, well, what's the mass properties of this thing? Could it be handled by some sort of handling device by hand? Attached to the robot arm? And then if we had to push it away, what kind of forces would we have to push on it to make it stable, and is that a reasonable thing to do? So we calculated a twenty- or thirty-pound push would be enough."

The calculations were right on. Lounge said computer simulations calculated a push of 27.36 pounds would be needed. Then the question was whether the rescue mission was even feasible. After looking into the challenges further, Lounge and van Hoften believed it was and encouraged their commander to seek approval for his crew to do the job. Joe Engle shepherded the request through center management and up to NASA headquarters and got the go-ahead for the recovery. Explained Lounge, "The key to the success of that mission and being able to do that was NASA was so busy flying shuttle missions that year that nobody was paying attention. If we'd had more attention, there'd have been a hundred people telling us why it wouldn't work and it's too much risk. But fortunately, there was a twelve-month period we flew ten missions; we were one of those."

Discussions about the feasibility of the SYNCOM recovery naturally led to comparisons with an earlier flight. One of the questions, recalled 51I pilot Dick Covey, was, would the astronauts be able to stop the rotation of the spacecraft? Covey said crewmate Ox van Hoften drew the solution on the back of a piece of paper.

He says, "It's only going to take this much force to stop the rotation, so that's not an issue." Then [they] said, "Well, you know, does anybody think that we could have a person stop the rotation and do that?" Ox says, "Well, here. Here's me," and he draws this big guy, and he says, "Here's the SYNCOM." He draws a little guy, and he says, "Here's Joe Allen, and there's a PALAPA [satellite]. So if he can grab that one, then I can grab this one." We said, "Okay, yes." It was the "big astronaut, little astronaut" approach to things.

For Commander Engle, this mission would be very different from STS-2, on which he also served as commander. Engle described his second shuttle flight as less demanding than his first; the biggest difference between the two flights, he said, was that on his second mission—and NASA's twentieth—there were more people there to help out.

We had only a crew of two on STS-2, and one of the lessons we learned from those first four orbital flight tests was that the shuttle—the orbiter itself—probably represents more of a workload than should be put onto a crew of two. It's just too demanding as far as configuring all of the systems and switches, circuit breakers. There are over fifteen hundred switches and circuit breakers that potentially have to be configured during flight, and some of those are in fairly time-critical times. . . . Some of them are on the mid-deck and some are on the flight deck, so you're going back and forth and around. Having more people on board really reduces the workload of actually flying the vehicle.

In flight, Engle discovered that having more crew to do the work meant more time personally to look out the window at Earth. "It's a very, very inspiring experience to see how thin, how delicate the atmosphere is, how beautiful the Earth is, really, what a beautiful piece of work it is, and to see the features go by," Engle described.

Sultan Al-Saud was assigned to our crew initially, when one of our payload satellites was the ARABSAT. [Al-Saud flew as part of the 51G crew.] He was assigned as a mission specialist on our crew, and when he eventually did fly, I think he said it better than anybody has. He said, "The first day or two in space, we were looking for our countries. Then the next day or two, we were looking at our continents. By about the fourth or fifth day, we were all looking at our world." Boy, it's one of those things that I said, "God, I wish I'd have thought of that. I wish I'd have said that."

The crew made several launch attempts before finally getting off the ground on 27 August 1985. The first launch attempt, on 24 August, was scrubbed at T minus five minutes due to thunderstorms in the vicinity.

"When we got back in the crew quarters after that first scrub for weather," recalled Covey,

[there was] John Young, who was the chief of the Astronaut Office at the time and also served as the airborne weather caller in the Shuttle Training Aircraft. . . . I was making some comments about, "I can't believe we scrubbed for those two little showers out there. Anybody with half a lick of sense would have said, 'Let's go. This could be a lot worse.'" John Young came over and looked at me and he says, "The crew cannot make the call on the weather. They do not

know what's going on. All they can see is out the window. That's other people's job." I said, "Yes, sir. Okay."

Launch was pushed to the next day, 25 August. Engle's birthday was 26 August, and on board the shuttle was a cake that the crew was taking into orbit to celebrate. The launch scrubbed again on the twenty-fifth, this time because of a failure with one of the orbiter's on-board computers, and was pushed to the twenty-seventh. Said Covey, "They wound up unstowing the birthday cake and taking it back to crew quarters, and we had it there instead of on orbit."

Initially, things weren't looking good for the third launch attempt, either, Engle said. It was raining so hard the crew wore big yellow rain slickers from the crew quarters to the Astrovan and up the elevator to the white room. Engle admitted that as the crew boarded, the astronauts "didn't think there was a prayer" of actually flying that day. But they had only one more delay before they would have to detank and refuel the shuttle, which would delay the mission an additional two days, so the decision was made to get ready and see what happened.

We got in the bird and we strapped in and we started countdown. Ox van Hoften was in the number-four seat, over on the right-hand side aft, and Mike Lounge was in the center seat aft, and we were sitting there waiting, and launch control had called several holds. Ox was so big that he hung out over the seats as he sat back, and he was very uncomfortable, and he talked Mike into unstrapping and going down to the mid-deck so that he could stretch across both those seats in the back of the flight deck. We were lying there waiting, and it was raining, and raining fairly good.

As the countdown clock ticked down nearer and nearer to the scheduled launch time, the crew continued to believe that there was essentially no chance of a launch that day.

We got down to five minutes or six minutes, and . . . we got the call from launch control to start the APUs. Dick Covey and I looked at each other kind of incredulously and asked them to repeat. And they said, "Start the APUs. We don't have much time in the window here." So he started going through the procedures to start the APUs, and they make kind of a whining noise as they come up to speed. The rest of the crew was asleep down in the mid-deck.

I think it was Fish [Bill Fisher] woke up and said, "What's that noise? What's going on?" We said, "We're cranking APUs. Let's go," or something like that. Dick was into the second APU, and they looked up and saw the rain coming down and they said, "Yeah, sure, we're not going anywhere today. Why [are] you starting APUs?" We didn't have time to explain to them, because the sequence gets pretty rushed then. So we yelled to them, "Damn it, we're going. We're going to launch. Get back in your seats and get strapped in." They woke up Ox and Mike, and they got back in their seats, and they had to strap themselves in. Normally you have a crew strap you in; they had to strap each other in. And Dick and I were busy getting systems up to speed and running, and all we could hear was Mike and Ox back there yelling at each other to, "Get that strap for me. Where's my comm lead?" "Get it yourself. I can't find mine." And they were trying to strap themselves in, and we were counting down to launch. They really didn't believe we were going to launch because it was, in fact, raining, but they counted right down to the launch and we did go. It went right through a light rain, but it was raining.

Engle said that after the crew returned to Earth at the end of the mission he asked about the decision to launch through the rain. It turns out that the weather spotters were flying at the Shuttle Launch Facility at launch time, and it was clear there. "[They asked,] 'Why didn't you tell us it was raining [at the pad]?' We used their rationale then. We said, 'Our job is to be ready to fly. You guys tell us when the weather's okay.'"

In addition to capturing and repairing the SYNCOM that malfunctioned on 51D, the mission was to deploy three other satellites. One of those was another SYNCOM, known as SYNCOM IV-4 or LEASAT-4; the other two were ASC-1, launched for American Satellite Company, and the Australian communications satellite AUSSAT-1.

"We were supposed to do one the first day, one the second day, one the third day, and then the fourth and fifth days were repair days, and there was a day in between," explained Lounge. But first, the crew was tasked with using the camera to look at the payload bay and the sun shield to make sure everything was intact after launch. Lounge did just that.

Then I commanded the sun shield open, and I had failed to stow the camera. If it had been day two instead of day one, I'd have been more aware of it. On day one you're just kind of overwhelmed and you're just down doing the steps, and it's not a good defense, but that was an example of why you don't change things at the last minute and why you don't do things you haven't simulated, because we'd never simulated that. That was some engineer or program manager said, "Wouldn't it be nice to add this camera task." Now I had a camera out of position, opened the sun shield against the camera, and it bent the sun shield and it got hung up on the top of the shuttle.

To address the problem, the crew did an earlier than planned checkout of the robotic arm and then used the arm to essentially bang against the sunshade, Lounge said. "[The sunshade was a] very flimsy structure with [an] aluminum tube frame and Mylar fabric, so not a lot to it, but it had to get out of the way."

Lounge maneuvered the arm, but it wasn't working right either.

The elbow joint had a problem that wouldn't let the automatic control system operate the arm, so I had to command the arm single-joint mode, which means instead of some coordinated motion, command the tip to move in a certain trajectory, you just had to say, all right, elbow, move like this; wrist, move like this, rotate like this. So, a little awkward and took awhile, but I got the arm down there and banged on the solar array and got it down, and then we deployed that one [satellite]. . . . We deployed both of them on the same day, five or six hours after launch. So that was exciting, more exciting than it needed to be.

After the satellite deployments came the SYNCOM repair attempt. The shuttle rendezvoused with the failed satellite, and robot arm operator Mike Lounge helped Bill Fisher and Ox van Hoften get ready for their EVA. Once the EVA began, van Hoften installed a foot platform on the end of the arm, and Lounge moved him toward the satellite so that he could grab it, just like in his napkin drawings months earlier.

The satellite, though, was in a tumble, so there were concerns about whether the capture could go as planned. As on earlier missions, the astronauts took advantage of the long periods of loss of signal and worked out a solution in real time.

"When we got up there and it was tumbling," said Covey,

we were trying to relay back to the ground what was going on. . . . We were trying to figure out stuff. They were trying to figure out stuff. Finally we went LOS and Ox said, "Fly me up to it," and he went up and he just grabbed it. If the ground would have been watching, we wouldn't have done that, I'm sure, like that. But he grabs it and spins it, just with his hands on the edge, where they say, "Watch out for the sharp edges." And he spins it a little bit so that the fixtures come around to him, and then he rotates it a little bit, and he gets that tool on, screws it down. We maneuver it down. We come AOS. *We say, "Well, we got it, Houston." They didn't ask why. They didn't ask how.*

STS-61B

Crew: Commander Brewster Shaw, Pilot Bryan O'Connor, Mission Specialists Mary Cleave, Jerry Ross, and Sherwood Spring, Payload Specialists Rodolfo Neri Vela (Mexico) and Charlie Walker

Orbiter: *Atlantis*

Launched: 26 November 1985

Landed: 3 December 1985

Mission: Deployment of three satellites, demonstration of space assembly techniques

As NASA worked to create a healthy manifest of shuttle flights, glitches with satellites' inertial upper stages and with payloads themselves made for an ever-shifting manifest. "Continuously, we were juggling the manifest," said astronaut Jerry Ross. "Crews were getting shifted from flight to flight. The payloads were getting shifted from flight to flight. And basically, throughout 1985, our crew trained for every mission that flew that year except for military or Spacelab missions."

At one point, Ross said, the 61B crew was even assigned to 51L, the ill-fated final launch of the Space Shuttle *Challenger*. After bouncing around to several different mission possibilities, the crew settled in on 61B.

The mission included two payload specialists, Charlie Walker and Rodolfo Neri Vela of Mexico. Ross remembered that the agency was being pressured to fly civilians—teachers, politicians, and the like. "We were giving away seats, is the way we kind of saw it, to nonprofessional astronauts, when we thought that the astronauts could do the jobs if properly trained," Ross said.

The flight was Walker's third flight in fifteen months and it was the first flight for a payload specialist from Mexico. "The guys did a great job on orbit," Ross praised the mission's two payload specialists. "They were always very helpful. They knew that if the operations on the flight deck were very hectic, they stayed out of the way, which is the right thing to do, frankly.

But at other times they would come up onto the flight deck and enjoy the view as well as any of the rest of the crew."

The mission deployed three more communications satellites: one for Mexico, one for Australia, and one for RCA Americom. All three satellites were deployed using Payload Assist Modules. Ross and Mission Specialist Sherwood Spring worked together on the deployments. The two also journeyed outside the shuttle on two spacewalks to experiment further with assembling erectable structures in space. The two experiments were the Experimental Assembly of Structures in Extravehicular Activity (EASE) and the Assembly Concept for Construction of Erectable Space Structure (ACCESS).

"I'll remember the day forever, when I got to go do my first spacewalk," recalled Ross, who throughout his career ventured out of his spacecraft for a total of nine EVAs. That first venture outside was something he had been looking forward to for quite some time.

I got a chance to do a lot of [support for] spacewalks as a CapCom on the ground, and I got a little bit more green with envy every time I did that, thinking about what those guys were doing, how much fun they were having. So when I ultimately got a chance to go outside for my first time, I was worried, because I was worried that the orbiter was going to have a problem, we were going to have to go home early, or one of the spacesuits wouldn't check out and we wouldn't be able to go out, and all those things.

I'll never forget opening up the hatch and poking my head out the first time, and I literally had this very strong desire to let out this war whoop of glee and excitement. But I figured that if I did that, they'd say, "Okay, Ross has finally lost it. Let's get his butt back inside," and that would have been it. But it felt totally natural, just totally natural to be outside in your own little cocoon, your own little spacecraft, and I felt basically instantly at home in terms of going to work.

EASE and ACCESS were designed to test how easily—or not—astronauts could assemble or deploy components in orbit. The idea was to study the feasibility of packing space structure truss components in a low-volume manner for transport to orbit so that they could then be expanded by astronauts in space. The question was more than just hypothetical; planning was already underway to construct a space station.

The idea of a space station was an old one, and NASA had already built and flown one space station, Skylab. When the Space Shuttle was proposed

30. Jerry Ross on a spacewalk demonstrating the first construction of large structures in weightlessness. Courtesy NASA.

in the late 1960s, NASA's desire was to build both the shuttle and a space station as the first steps in developing an infrastructure for interplanetary exploration. The administration of President Richard Nixon approved funding only for the Space Shuttle, however. By 1984, Reagan believed that the Space Shuttle program was sufficiently mature to move ahead with a space station, and the Space Station *Freedom* project was born. The station had ambitious goals, with plans calling for it to be a microgravity science lab, a repair shop for satellites, an assembly port for deep-space vehicles, and a commercial microgravity factory. Unlike Skylab, which was launched fully assembled atop a powerful Saturn V rocket, plans were for *Freedom* to be launched in multiple modules aboard the shuttle and assembled in orbit. The 61B spacewalks to test EASE and ACCESS were part of the preparations, designed to study how best to build components so that they could be flown compactly and easily assembled by astronauts in space.

"The second spacewalk, we worked off the end of the mechanical arm for a lot of the work," Ross explained.

We did the assembly, the top bay of the ACCESS truss off the end of the arm. We simulated the running of the electrical cable. We did the simulation of doing a repair of the truss by taking out and reinserting an element there. We removed the trusses off of the fixture and maneuvered them around to see how that would be in terms of assembling a larger structure. We also mounted a U.S. flag that we had modified onto the truss and took some great pictures of us saluting the flag on the end of the arm up there. We also made a flag that we took outside. We called ourselves the Ace Construction Company. There's a series of Ace signs that were taken outside on various spacewalks. . . . Somehow we've lost some of that fun over the years. I'm not sure why.

For the ACCESS work, the RMS arm, with the spacewalkers at its end, was operated by Mary Cleave, whose height, or lack thereof, required special accommodation for her to be able to perform the task, according to Commander Shaw.

In order for her to get up and be able to look out the window and operate the controls on the RMS, we'd strung a bungee across the panel and she'd stick her legs in front of that bungee and it would hold her against the panel . . . so she could be high enough and see and be in the right position to operate the RMS. I remember coming up behind Mary once when she was operating the RMS and there was somebody on the end of the arm. I put my hand on her shoulder, and her whole body was quivering, because she was so intent on doing this job right and not hurting anybody, and so focused and so conscientious, not wanting to do anything wrong, because she knew she had somebody out there on the end of this arm, and she was just quivering, and that just impressed the hell out of me, because I thought, you know, what a challenge, what a task for her to buy into doing when it obviously stressed her so.

All in all, Ross said, the two experiments were successful in their goal of producing data about in-space construction. "It gave us quite a bit of understanding and knowledge of what it would be like to assemble things in space," Ross noted.

Ultimately, that's not the way that we chose to build the station, because when you think about having to integrate all the electrical and fluid lines and everything else into the structure once you've assembled this open network of truss, it becomes harder to figure out how you're going to do that and properly connect everything together and make sure it's tested and works properly. But we did learn a lot about assembling things in space and proved that they are valid things that you could anticipate doing, even on the current station, at some point, if you needed to add a new antenna or something like that.

One advancement that came out of the ACCESS and EASE experiments wasn't even in space but had a big impact down the road. NASA realized through training for the space assembly tests that the Weightless Environmental Training Facility, or WET-F, water tank where astronauts were training for EVA was not going to be large enough to train for construction of a large space station. "[In] the facility we had when we built the ACCESS truss, we could only build like one and a half bays before it started sticking out of the surface of the water," recalled Ross. "And the EASE experiment, when we did it, basically our backpacks of our suits when we were at the top of the structure were right at the surface of the water. So if you're going to build anything that's anywhere close to being big on orbit, that wasn't going to get it."

For the next ten years, Ross helped NASA campaign to Congress for funds for a new facility. Ross helped design the requirements for the facility and led the Operational Readiness Inspection Team that eventually certified the new Neutral Buoyancy Laboratory at the Sonny Carter Training Facility.

During one of the two ACCESS/EASE spacewalks, Ross recalled saying to Spring, "Let's go build a space station." Ross would later have the opportunity to repeat that same phrase on his final spacewalk, on STS-110 during actual assembly of the International Space Station in 2002.

The crew of 61B was the first to be on the shuttle in orbit on Thanksgiving Day, which meant, of course, that the astronauts needed a space-compatible Thanksgiving feast. Payload Specialist Charlie Walker recalled that the crew worked with JSC foods manager Rita Rapp on planning a special meal for the holiday. Rapp had been involved in space food development and astronaut menus since the early Mercury missions. Walker said the crew specifically requested pumpkin pie.

Of course, the menu had to be approved by NASA to withstand launch, and pumpkin pie didn't make the cut. "Apparently somebody did the jiggle test, the vibration test, on the pumpkin pie, and what we were told later was, 'Well, pumpkin pie does not make it to orbit. The center of the pumpkin pie turns back to liquid, so you won't have pumpkin pie, you'll have pumpkin slop in orbit, and you really don't want that, so sorry, no pumpkin pie.' So Rita said, 'All right, how about pumpkin bread? We can do that, and that will work, we know.' So we had pumpkin bread on orbit for Thanksgiving."

Thanksgiving was not the only interesting part of the mission from a culinary perspective. Being from Mexico, Payload Specialist Rodolfo Vela brought into space with him foods from home, one of which significantly changed the way astronauts would eat from that point forward. "Rodolfo had, of course, the desire, and probably the need, as it was perceived back home from Mexico, to be seen to be flying with some local Mexican cultural things, and so food was one of those," Walker said.

One of the things that Rodolfo wanted to fly with, of course, was flour tortillas. In retrospect, I think that this amounted to something of a minor revolution in the U.S. manned space program, in that up to that time, of course, when crews wanted to have sandwiches in orbit, well, you went into the pantry, and you took out the sliced bread, sliced leavened bread that had been flown, for your sandwiches. Well, sliced bread, of course, always results in some degree of crumbs, and the crumbs don't fall to the floor in the cabin in space. They are all around you, in your eyes, in your hair, and so it's messy and just not that attractive. The crew saw Rodolfo flying with these flour tortillas and immediately thought, "Ooh, this may be real good," and it was real good. It was tasty, after all, but when you took spread or anything that you wanted to make into a rolled sandwich and devoured it that way, but it was just no-muss, no-fuss kind of thing. I remember taking some sliced bread, but there may have been some sliced bread that even made it home, because we just found that the flour tortilla thing was well in advance of sliced bread, crumbly bread, for the preparation of sandwiches or just as a bread to go with your meal. The flour tortillas worked well, much better than that.

Pilot Bryan O'Connor played a prank on Mission Specialist Spring. Spring was in the army—a West Point grad—and O'Connor was from the

Naval Academy. During the mission the two armed forces faced off in the annual Army-Navy Game. O'Connor's prank centered around the rivalry between the two forces.

Each person was allowed to carry six audios, and NASA *would help you record records or whatever you wanted onto these space-qualified audiotapes. Then we would carry them and a tape player on board with our equipment. Usually what would happen is people would break those out when it was time for bed and listen to their favorite music at bedtime. . . . It was on day three that we turned off the lights and, I don't know, it was about ten minutes after the lights were off, and I was borderline asleep, and I hear this loud cry from the other side of the mid-deck, where Woody [Spring] was hanging off the wall in his bed. He yells out, "O'Connor, you S.O.B.!" It woke me up with a start, and I had no idea what he was talking about. "What is it? What is it?" He says, "You know what it is."*

And all of a sudden, it clicked with me. About a month before flight, when we were having the people transcribe music onto our tapes, I went over to the guy that was working on Woody's tapes and I gave him a record with the Naval Academy fight song on it and I said, "I want you to go right in the middle of his tape somewhere, just right in the middle, and superimpose the navy fight song somewhere on his tape." Well, it turned out it was his Peter, Paul, and Mary album, and it was right in the middle of "I've Got a Hammer." He's listening to "I've Got a Hammer" on his way to sleep and suddenly up comes this really loud navy fight song thing right in the middle of it. We still joke about that to this day. In fact, sometimes we go to one or the other house and watch the Army-Navy Game together, and we always remember that night on the Atlantis *in the mid-deck.*

The mission ended and the crew came in for landing at Edwards Air Force Base in California. Weather brought the crew in one orbit sooner than was originally planned. "We came to wheels stop, and everybody unbuckles, and they're trying to get their land legs again," said Walker.

Jerry [Ross] is over at the hatch real quickly and wants to pop the hatch open so that we've got that part of the job done. Well, Jerry pops the hatch open, but it literally pops open, because whomever had planned these things had forgotten about the altitude, pressure altitude difference, between sea level at the Cape and the probably three-thousand-, four-thousand-foot elevation at Edwards Air Force Base. So it's a little bit less air pressure outside. Well, we're still at sea-level pressure inside the ship. So he turns the crank on the side hatch, and the hatch goes, "Pow!" It flops down, and right away, I think Jerry said something about, "Oh, my God, I'm going to have to pay for a new hatch."

STS-61C
Crew: Commander Hoot Gibson, Pilot Charles Bolden, Mission Specialists Franklin Chang-Diaz, Steven Hawley, and Pinky Nelson, Payload Specialists Robert Cenker and Congressman Bill Nelson
Orbiter: *Columbia*
Launched: 12 January 1986
Landed: 18 January 1986
Mission: Deployment of the RCA satellite SATCOM KU-1, various other experiments

The payload specialists on 61C were Robert Cenker of RCA, who during the mission observed the deployment of the RCA satellite, performed a variety of physiological tests, and operated an infrared imaging camera; and the second member of the U.S. Congress to fly, Bill Nelson, a member of the House of Representatives, representing Florida and its Space Coast, and chairman of the Space Subcommittee of the Science, Space, and Technology Committee.

Pinky Nelson recalled that the payload specialists assigned to the mission changed several times leading up to flight. "Our original payload specialists were Bob Cenker and Greg Jarvis, so they were training with us," Pinky said.

It was after Jake Garn flew, and then they decided they had to offer a flight to his counterpart in the House. Don Fuqua couldn't fly for some reason, and so it filtered down to the chair of the subcommittee, Bill Nelson, and he jumped at the chance. Who could blame him? This was just months before the flight, in the fall or late fall, even. The flight was scheduled in December. So they bumped Greg and his little payload off the mission over onto Dick Scobee's [51L Challenger*] crew and added Bill to our crew. I think our attitude generally at that point was, "Well, that's just the way the program's going. We're flying payload specialists. We'll make the best of it."*

Pinky Nelson described Representative Nelson as a model payload specialist, working very hard to contribute to the mission. "He had no experience either in aviation or anything technical. He was a lawyer, so he had a huge learning curve, but that didn't stop him from trying, and I think he knew where his limitations were," Pinky said. "He wanted to jump in and help a lot of times, but just didn't have the wherewithal to do it, but worked very hard and was incredibly enthusiastic."

The launch of 61C was delayed seven times. Originally set for 18 December 1985, the launch was delayed one day when additional time was needed to close out the orbiter's aft compartment. On 19 December the launch scrubbed at T minus fourteen seconds due to a problem with the right solid rocket booster hydraulic power unit. "As it turned out," Charlie Bolden said, "when we finally got out of the vehicle and they detanked and went in, they determined that there wasn't really a problem. . . . It was a computer problem, not a physical problem with the hydraulic power unit at all, and it probably would have functioned perfectly normally, and we'd have had a great flight."

The launch was pushed out eighteen days, to 6 January 1986. The third attempt stopped at T minus thirty-one seconds due to the accidental draining of approximately four thousand pounds of liquid oxygen from the external tank. The next day launch scrubbed at T minus nine minutes due to bad weather at both transoceanic abort landing sites.

Two days later, on 9 January, launch was delayed yet again, this time because a launchpad liquid oxygen sensor broke off and lodged in the number two main engine prevalve. "That time we got down to thirty-one seconds, and one more time things weren't right," Bolden said. "So we got out, and it was another main engine valve. This time they found it. There had actually been a probe, a temperature probe, that in the defueling, they had broken the temperature probe off, and it had lodged inside the valve, keeping the valve from closing fully. So that would have been a bad day. That would have been a catastrophic day, because the engine would have exploded had we launched."

A 10 January launch was delayed two days due to heavy rains. After so many delays, despite the adverse weather conditions, the crew was still loaded onto the vehicle, on the off chance that things would happen to clear up. "It was the worst thunderstorm I'd ever been in," Bolden said.

We were really not happy about being there, because you could hear the lightning. You could hear stuff crackling in the headset. You know, you're sitting out there on the top of two million pounds of liquid hydrogen and liquid oxygen and two solid rocket boosters, and they told you about this umbrella that's over the pad, that keeps lightning from getting down there, but we had seen lightning actually hit the lightning-arrester system on STS*-8, which was right there on the launchpad. So none of us were enamored with being out there, and we started talking about the fact that we really ought not be out here.*

While some astronauts found multiple scrubs and launch attempts frustrating, Steve Hawley said he wasn't bothered by it. "My approach to that has always been, hey, you know, I'd go out to the launchpad every time expecting not to launch," Hawley said.

If you think about all the things that have to work, including the weather at several different locations around the world, in order to make a launch happen, you would probably conclude, based on the numbers, that it's not even worth trying. So I always figured that we're going to turn around and come back. So I'm always surprised when we launch. So my mindset was always, you know, we'll go out there and try and see what happens. So I never really viewed it as a disappointment or anything. I always feel a little bad for, you know, maybe family and guests that may have come out to watch, that now they have to deal with the fact there's a delay and whether they can stay, whether they have to leave, and that's kind of a hassle for them, but it never bothered me particularly.

In those days, Hawley pointed out, the launch windows were much longer than they were later in the program for International Space Station docking missions. In the early days launch windows were two and a half hours, versus five to ten minutes for ISS flights. The longer windows were advantageous in terms of probability of launch but could be adverse in terms of crew comfort. "You're out there on your back for five or six hours, and that gets to be pretty long, day after day, but the fact that you didn't launch never bothered me particularly."

Having been part of multiple scrubs on his earlier 41D mission, Hawley had a reputation for not being able to launch. "I don't remember how we came up with the specifics of the disguise, but I decided that if it didn't know it was me, then maybe we'd launch, and so I taped over my name tag with gray tape and had the glasses-nose-mustache disguise and wore that into the [white] room. I had the commander's permission, but I don't remember if we had told anybody else we were going to do that. . . . Evidently it worked, because we did launch that day."

So finally, on 12 January 1986 the mission made it off the ground. By this time the crew members were wondering if they'd ever really go. "We did all kinds of crazy stuff," Bolden said, "fully expecting that we wouldn't launch, because I think the weatherman had given us a less than 50 percent chance that the winds were going to be good or something, so we went out and we were about as loose as you could be that morning. And they went through the countdown, came out of the holds and nothing happened. 'Ten, nine, eight, seven, six.' And we looked at each other and went, 'Holy—we're really going to go. We'd better get ready.' And the vehicle started shaking and stuff, and we were gone."

Within seconds of lifting off, an alarm sounded. "I looked down at what I could see, with everything shaking and vibrating, and we had an indication that we had a helium leak in—I think it was the right-hand main engine," Pilot Bolden recalled. "Had it been true, it was going to be a bad day. . . . We didn't have a real problem. We had a problem, but it was an instrumentation problem."

With the determination made that the indicator was giving a false reading, the crew continued on into space. "We got on orbit and it was awesome," said Bolden of his first spaceflight. "It was unlike anything I'd expected. Technically, we were fully qualified, fully ready, and everything. Emotionally, I wasn't even close. I started crying. Not bawling or anything, but just kind of tears rolling down my cheek when I looked out the window and saw the continent of Africa coming up. It looked like a big island. Just awesome, unlike anything I'd ever imagined."

Pinky Nelson said he found the entire mission rather frustrating, describing the mission as trivial. "We launched one satellite, and we did this silly material science experiment out in the payload bay which didn't work. I knew it wasn't going to work when we launched. Halley's Comet was up at the time and we had this little astronomy thing to look at Halley's Comet, and it was launched broken, so it never worked. So the mission itself, to say what we did, I don't know. I deployed a satellite.

Steve deployed a satellite. We threw a bunch of switches, took a bunch of pictures."

According to Bolden, the crew referred to its mission as an end-of-year-clearance flight. "We had picked up just tons of payload, science payloads that Marshall [Space Flight Center, in Huntsville, Alabama] had been trying to fly for years, and some of the Spacelab experiments and stuff that they couldn't get flown," Bolden said. "So we had Congressman Nelson and every experiment known to man that they couldn't get in. There was nothing spectacular about our mission. It was almost like a year-end clearance sale."

Hawley said there was a general feeling at the time that without Nelson on the flight the mission might have been canceled. "We all suspected, although no one ever said, that because of the delays that we got into and the fact that, frankly, our payload wasn't very robust, that were it not for his presence on the flight, we might have been canceled. . . . We wondered about that and always thought, without knowing for sure, that that might have happened if we hadn't had a congressman, but this was his flight, and so we had some guarantee that it would happen."

In addition to launching the RCA satellite SATCOM KU-1, the mission was assigned the Comet Halley Active Monitoring Program (CHAMP) experiment, which was a 35 mm camera that was to be used to photograph Halley's Comet. But the camera didn't work properly due to battery problems. Science experiments on board were the Materials Science Laboratory-2, Hitchhiker G-1, Infrared Imaging experiment, Initial Blood Storage experiment, Handheld Protein Crystal Growth experiment, three Shuttle Student Involvement Program experiments, and thirteen Getaway Specials.

"The big challenge," Bolden said,

was arguing with the ground about how we should do some of the experiments. There were some that we could see were not going exactly right. I didn't have the problem as much as Pinky. Pinky was the big person working a lot of the material sciences experiments. And while we had very little insight into what was going on inside the box, we could tell that because of the data that we were seeing on board, we could tell that if we were just given an opportunity to reenergize an experiment, or to turn the orbiter a different way, or do something, we might be able to get some more data for the principal investigators. The principal investigators agreed, but the flight control crew on the ground [said] that wasn't in the plan. They weren't interested in ad-libbing. They had a flight to fly and a plan to fly, and so forget about these doggone experiments.

Bolden said there was generally always a power struggle between the flight director on the ground and the commander of the shuttle in space, an issue that dated back to the earliest U.S. spaceflights.

There's always a pull and tug between the flight director, who is in charge—nobody argues that point—and the crew commander, who, by the General Prudential Rule of Seamanship of navigation at sea, has ultimate responsibility for the safety of the crew and vessel. If there's a disagreement between the commander and the flight director—and that's happened on very, very few occasions, but every once in a while it happens—the commander can do what he or she thinks is the right thing to do and is justified in doing that by the General Prudential Rule. And even NASA recognizes that. Now, you could be in deep yogurt when you come back, if something goes wrong. But you have the right to countermand the direction of the flight director. Almost never happens.

Bolden recalled that because the agency wanted to get *Challenger* and the Teacher in Space mission off the ground, 61C was cut from six days to four. The flight was scheduled to land at Kennedy Space Center—which would have made it the first flight to land there since the blown-tire incident on 51D—but the weather in Florida once again didn't cooperate.

"They kept waving us off and making us wait another day to try to get back into KSC," Hawley recalled. "What I remember is that by the third day we had sort of run out of most everything, including film, and part of our training had been to look for spiral eddies near the equator, because the theory was, for whatever reason, you didn't see them near the equator, and Charlie was looking out the window and claimed to see one, and I told him, 'Well, you'd better draw a picture of it, because we don't have any film.' So we couldn't take a photograph."

After two days of bad weather preventing the shuttle from landing in Florida, it was decided to send *Columbia* to Edwards Air Force Base. "The first attempt at Edwards was waved off because the weather there was bad," Bolden said. "Finally, on our fifth attempted landing . . . in the middle of the night on 18 January, we landed at Edwards Air Force Base, which was interesting because with a daytime scheduled landing, you would have thought that we wouldn't have been ready for that. And Hoot, in his infinite wisdom, had decided that half of our landing training was going to be nighttime, because you needed to be prepared for anything. And so we were as ready for a night landing as we could have been for anything."

The landing was smooth, but Bolden said Congressman Nelson was disappointed not to have landed in his home state. "He really had these visions of landing in Florida and taking a Florida orange or something, and boy, the crew that picked us up was unmerciful, because they came out with a big—it wasn't a bushel basket, it was a peck—basket of California oranges and grapefruits. And even having come from space, he was just not in a good mood. So that was a joke that he really did not appreciate."

Pinky Nelson recalled that Bill Nelson was struggling with not feeling well after landing but kept going anyway. "Most people don't feel very good their first day or two in space, but don't have too much trouble when they get back on the ground," Pinky said. "Bill had a really hard time for a few hours after we landed, but boy, he was a trooper. He was suffering, but he—you know, good politician—put on a good face, and we had to do our little thing out at Edwards and all that and get back on the plane. He really sucked it up and hung in there, even though he was barely standing. And the rest of us, of course, cut him no slack at all."

Bolden described being in awe of Gibson's skills as a pilot and told how the five-time shuttle veteran took Bolden under his wing. "The way that I was trained with Hoot was you don't ever wing anything," Bolden said.

I credit him with my technique as a commander. He preached from day one, "We don't ever do anything from memory. We don't ever wing it. If something's going to happen, there is a procedure for it. And if there's not a procedure for it, then we're going to ask somebody, because somebody should have thought about it." And so what we did was we trained ourselves just to know where to go in the book. And hopefully, crews still train like that, although I always flew with people who would invariably want to wing it, because they prided themselves in having photographic memories or stuff like that. The orbiter and just spaceflight is too critical to rely on memory, when you've got all of these procedures that you can use, and the ground to talk to.

Gibson taught him what Bolden referred to as Hoot's Law, a piece of wisdom that has stayed with him ever since. "We were in the simulator one day," Bolden said, "in the SMS [Shuttle Motion Simulator], and I was still struggling. It was in my struggling phase. And I really wanted to impress everybody on the crew and the training team. We had an engine go out, boom, like that, right on liftoff."

During training, Bolden explained, the sim crew would introduce some errors to see how the astronauts would respond to them, but then at other times, they introduced abnormalities simply to try to distract the astronauts from the important things. "There is probably one critical thing that you really need to focus on, and the rest of it doesn't make any difference. If you don't work on it, you get to orbit and you don't even know it was there. But if you notice it and start thinking about it or start working on it, you can get yourself in all kinds of trouble. They love doing that with electrical systems, so they would give you an electrical failure of some type."

Bolden recalled that on this particular occasion he was working an engine issue, and the sim team introduced a minor electrical problem. He first made the mistake of trying to work the corrective procedure on the wrong electrical bus, one of multiple duplicate systems on the shuttle.

The training team intended it this way. You learn a lesson from it. So I started working this procedure and what I did in safeing the bus was I shut down the bus for an operating engine. When I did that, the engine lost power and it got real quiet. So we went from having one engine down in the orbiter, which we could have gotten out of, to having two engines down, and we were in the water, dead. Here I went from I was going to feel real good about myself because I'd impress my crew to feeling just horrible because I had killed us all. And Hoot kind of reached over and patted me on the shoulder. He said, "Charles, let me tell you about Hoot's Law." That's the way he used to do stuff sometimes. And I said, "What's Hoot's Law?" And he said, "No matter how bad things get, you can always make them worse."

And I remembered Hoot's Law from that day. That was probably 1984, or 1985 at the latest, early in my training. But I remembered Hoot's Law every day. I have remembered Hoot's Law every day of my life since then. And I've had some bad things go wrong with me in airplanes and other places, but Hoot's Law has always caused me to take a deep breath and wait and think about it and then make sure that somebody else sees the same thing I did. And that's the way I trained my crews, but that was because of that experience I had with Hoot.

STS-62A

Crew: Commander Bob Crippen, Pilot Guy Gardner, Mission Specialists Dale Gardner, Jerry Ross, and Mike Mullane, Payload Specialists Brett Watterson and Pete Aldridge

Orbiter: *Discovery*

Launched: N/A

Landed: N/A

Mission: Deployment of reconnaissance satellite, first launch from Vandenberg

As 1985 wrapped up and 1986 began, the Space Shuttle was beginning to realize its promised potential. The early flights had hinted at that promise, but in 1985, with the orbiter's high flight rate, variety of payloads, and distinguished payloads, the nation was beginning to see that potential become reality. And as those flights were taking place, work on the ground was foreshadowing even greater times ahead—an even greater flight rate, a teacher and journalist flying into space, the imminent launch of the Hubble Space Telescope, and entirely new classes of missions in planning that would mark even broader utilization for America's versatile Space Transportation System.

After his completion of mission 41G, Bob Crippen sought and was given command of a groundbreaking mission, 62A. The mission was the first time that the "2" was used in a designation, indicating that the launch was to take place not out of Kennedy Space Center but rather out of Vandenberg Air Force Base in California.

"The air force built the launch site out there to do military missions which required a polar orbit, and it was a flight I wanted a lot," said Crippen, who explained that the assignment was a homecoming of sorts for him. During the 1960s, before his transfer to NASA, Crippen had been a part of the air force astronaut corps, based at Vandenberg and assigned to the Manned Orbiting Laboratory space station program. The MOL launches would have used Space Launch Complex 6 (SLC-6) at Vandenberg, and now SLC-6 had been modified to support Space Shuttle launches.

"I felt like I'd come full circle, and I really wanted that polar flight," Crippen said. "I lobbied for it and ended up being selected, although not without some consternation. I think since this was primarily an air force mission, there was a big push by the air force to have an air force commander on the flight. But the powers that be ended up discussing it a lot and letting me take the lead on it."

The crew, which included U.S. undersecretary of the air force Pete Aldridge, spent a lot of time at Vandenberg, Crippen said, making sure the launch complex was acceptable. "We actually took the *Enterprise* out there and used it to run through where they had to move it to stack it, and they actually had an external tank and some not-real solid rockets out there. . . . So we mounted it all up, and I've got pictures of the vehicle sitting on the launchpad like it's ready to fly, but it was the *Enterprise*, as opposed to the *Discovery*."

Jerry Ross recalled that a key difference between Kennedy and Vandenberg was that the shuttle stack was going to be assembled on the launchpad itself. At Kennedy, the orbiter, SRBs, and external tank were stacked inside the massive Vehicle Assembly Building and then moved to the launchpad on the crawler. "The solid rocket motors were going to be stacked up out on the pad. The external tank would be mounted to those out at the pad, and then the shuttle [orbiter] would be brought out on this multitiered carrier from its processing facility several miles away and taken out to the launchpad and put in place once everything else was ready," said Ross. "The entire launch stack could be enclosed in basically a rollaway hangar type of facility, and also the Launch Control Center was basically underneath the pad. It was buried in the concrete, not directly underneath, but still right there contiguous to the launchpad itself. That should have been a fairly noisy place to operate out of."

Ross learned that he had been assigned to 62A while still in training for 61B. "I was assigned to a second flight before I flew the first one," he said.

I was very excited about that, and the fact that you're going to get to do something so unique like that for the very first time was fascinating to me. When I launched in November of '85, I was supposed to fly again in January of '86, out of Vandenberg. Of course, everybody knew that date was not realistic at that point. But while I was on orbit, that date had been slipped out to July of '86, and most people thought that that was a fairly realistic date. So that would have been very close, two flights within six, seven months of each other.

Training for 61B kept him quite busy, but Ross managed to squeeze out a little extra time to also train with the 62A crew at Vandenberg. "I was worried with the flights getting so close together that maybe they were going to replace me," Ross admitted. "I talked to Crip a couple of times about that, and he said, 'Don't worry. We'll take care of you.'"

The planned launch, Ross said, would have been a fascinating ride. "We were going to go into a 72½ degree inclination orbit. . . . It would have been awesome. We'd have basically seen all the land masses of the world, so it would have been neat."

Fellow 62A crew member Mike Mullane agreed. "The idea of flying into polar orbit, oh, man, I was just looking forward to that so much. You're basically going to see the whole world. In an equatorial orbit like we were flying, or a low-inclination orbit like we were flying on the first mission, you don't get to see lot of the world. So I was really looking forward to that."

As with other DoD missions the shuttle had flown, preparations for 62A involved a high amount of secrecy, with astronauts required to not even tell their wives what they were doing. However, Mullane said he enjoyed working on military missions; his next two shuttle missions were also military missions. "You had a sense of this national security involved about it, which made you feel a little bit more pride, I guess, in what you were doing and importance in what you were doing."

Ross said the original flight plan had included twenty-four-hour-a-day operations by two air force payload specialists—Brett Watterson and Randy Odle—but Odle was bumped in favor of Pete Aldridge, undersecretary of the air force. "That would have been some pretty high-power folks flying with us on the flight."

The mission was assigned two main payloads: an experimental infrared telescope called Cirris and a prototype satellite, P-8888, called Teal Ruby. "My understanding was [Teal Ruby] was a staring mosaic infrared sensor satellite that was trying to be able to detect low-flying air-breathing vehicles, things like cruise missiles, and a way to try to detect those approaching U.S. territories."

Mullane said he had no additional concern at the prospect of launching from a previously unused launch complex. "Not any more beyond a natural terror of riding a rocket," said Mullane.

I don't care where it was launching from; I didn't personally have any fear about it being a new launchpad and therefore more danger associated with it. It's just that on launch on a shuttle, you fly with no escape system: no ejection seat, no pod, no parachute of any form. You fly in a rocket that has a flight-destruct system aboard it, so it can be blown up in case something goes wrong. Those are reasons why you're terrified. It's not where you're launching from, in my opinion; it's [that] the inherent act of flying one of these rockets is dangerous.

STS-61F
Crew: Commander Rick Hauck, Pilot Roy Bridges, Mission Specialists Mike Lounge and David Hilmers
Orbiter: *Challenger*
Launched: N/A
Landed: N/A
Mission: Launch of the *Ulysses* space probe

While Crippen and his crew were getting ready for 62A, Rick Hauck was preparing for his own first-of-its-kind Space Shuttle mission. He was the astronaut project officer for the Centaur cryogenically fueled upper stage, which NASA was planning to use on the shuttle as a platform for launching satellites.

The Centaur upper stage rocket had a thin aluminum skin and was pressure stabilized, such that if it wasn't pressurized, it would collapse under its own weight, like the Atlas missiles used to launch the orbital Mercury missions. "If it were not pressurized but suspended and you pushed on it with your finger, the tank walls would give and you'd see that you're flexing the metal," said Hauck. "Its advantage was that it carried liquid oxygen and liquid hydrogen, which, pound for pound, give better propulsion than a solid rocket motor [like NASA had been using on previous missions to boost satellites]."

Preparations were being made for the Centaur to launch two interplanetary probes—the *Ulysses* probe and the *Galileo* probe—which needed the powerful rockets available to be launched into deep space from the shuttle.

"At some point," Hauck said, "the decision was made, well, we've got to use the Centaur, which was never meant to be involved in human spaceflight." The origins of Centaur are older than NASA itself; it began as a project of the air force in 1957. Throughout its history it has been useful as an upper stage on expendable launch vehicles for launching satellites and probes to the moon and to planets other than Earth. But there was a high level of danger involved in pairing the highly volatile Centaur with a shuttle full of people.

"Rockets that are associated with human spaceflight have certain levels of redundancy and certain design specifications that are supposed to make them more reliable," Hauck commented.

Clearly, Centaur did not come from that heritage, so, number one, was that going to be an issue in itself, but number two is, if you've got a return-to-launch-site abort or a transatlantic abort and you've got to land, and you've got a rocket filled with liquid oxygen/liquid hydrogen in the cargo bay, you've got to get rid of the liquid oxygen and liquid hydrogen, so that means you've got to dump it while you're flying through this contingency abort. And to make sure that it can dump safely, you need to have redundant parallel dump valves, helium systems that control the dump valves, software that makes sure that contingencies can be taken care of. And then when you land, here you're sitting with the shuttle Centaur in the cargo bay that you haven't been able to dump all of it, so you're venting gaseous hydrogen out this side, gaseous oxygen out that side, and this is just not a good idea.

Hauck was working those issues when George Abbey called on him to command the first flight of the Shuttle-Centaur to launch the *Ulysses* solar probe. Astronaut Dave Walker was assigned the second Centaur mission, to launch the *Galileo* probe to Jupiter. The two missions had to be flown close together—in the first ten days of April 1986, Hauck said—because of the positioning of Earth in its orbit relative to the two satellites' destinations. "It was clear this would be very difficult," Hauck said. "We were going to have just four crew members, because that minimized the weight. We were going to 105 nautical mile altitude, which was lower than any shuttle had ever gone to, because you need the performance to get the Shuttle-Centaur up because it was so heavy."

The Shuttle-Centaur integration was being managed out of the Lewis Research Center, now Glenn Research Center, in Cleveland, Ohio, where the Centaur was developed originally. "Lewis had been the program managers for Atlas-Centaur, and so they knew the systems," Hauck said,

but in retrospect, the whole concept of taking something that was never designed to be part of the human spaceflight mission, that had this many potential failure modes, was not a good idea, because you're always saying, "Well, I don't want to solve the problems too exhaustively; I'd like to solve them just enough so that I've solved them." Well, what does that mean? You don't want to spend any more money than you have to, to solve the problem, so you're always trying to figure out, "Am I compromising too much or not?" And the net result is you're always compromising.

The head of the Office of Spaceflight at that time was Jess Moore, whom Hauck described as a good man but one who was unfamiliar with the world of human spaceflight. "Jess made it very clear that he wanted Dave and myself to be part of all the substantive discussions, and he was very sensitive to the human spaceflight issues, but he wasn't a human spaceflight guy," Hauck said. "I think that the program would have profited at that point by having had someone there who was more keenly attuned to the human spaceflight issues. As I say, he couldn't have been nicer to us and encouraged us more and bent over backwards to be sensitive to the issues, but he didn't start out as a human spaceflight guy."

In early January 1986, Hauck recalled, he worked on an issue with redundancy in the helium actuation system for the liquid oxygen and liquid hydrogen dump valves. It was clear, in Hauck's mind, that the program was willing to compromise on the margins in the propulsive force being provided by the pressurized helium, which concerned him enough that he took it up with Chief of the Astronaut Office John Young. "John Young called this mission 'Death Star,'" recalled Hauck. "That was his name for this mission, which he said with humor, but behind humor, there's a little bit of truth. I think it was conceded this was going to be the riskiest mission the shuttle would have flown up to that point."

Young, Hauck, and other members of the Astronaut Office argued before a NASA board why it was not a good idea to compromise on this feature, and the board turned down the request. "I went back to the crew office and I said to my crew, in essence, 'NASA is doing business differently from the way it has in the past. Safety is being compromised, and if any of you want to take yourself off this flight, I will support you.'"

Hauck said he didn't consider asking to be removed from the mission himself.

I probably had an ego tied up with it so much that, you know, "I can do this. Heck, I've flown off of aircraft carriers, and I've flown in combat, and I've put myself at risk in more ways than this, and I'm willing to do it." So I didn't ever think of saying, "Well, I'm not going to fly this mission." Knowing what I know now, with Challenger *and* Columbia, *maybe I would. But* NASA *was a lot different back there, when we'd never killed anybody in spaceflight up to that point. I mean, there was a certain amount of sense that it wouldn't happen.*

13. To Touch the Face of God

And, while with silent lifting mind I've trod
The high untrespassed sanctity of space,
Put out my hand and touched the face of God.

—*"High Flight," Pilot Officer Gillespie Magee, No. 412 Squadron, Royal Canadian Air Force, died 11 December 1941*

STS-51L
Crew: Commander Dick Scobee, Pilot Michael Smith, Mission Specialists Ellison Onizuka, Judy Resnik, and Ron McNair; Payload Specialists Christa McAuliffe and Gregory Jarvis
Orbiter: *Challenger*
Launched: 28 January 1986
Landed: N/A
Mission: Deployment of TDRS, astronomy research, Teacher in Space

Astronaut Dick Covey was the ascent CapCom for the 51L mission of the Space Shuttle *Challenger*. "There were two CapComs, the weather guy and the prime guy, and so it had been planned for some time that I'd be in the prime seat for [51L] and be the guy talking to them. . . . As the ascent CapCom you work so much with the crew that you have a lot of [connection]. In the training periods and stuff, not only do you sit over in the control center while they're doing ascents and talk to them, but you also go and work with them on other things."

Covey remembered getting together with the crew while the astronauts were in quarantine at JSC, before they flew down to Florida, to go over

31. Crew members of mission STS-51L stand in the White Room at Launchpad 39B. *Left to right*: Christa McAuliffe, Gregory Jarvis, Judy Resnik, Dick Scobee, Ronald McNair, Michael Smith, and Ellison Onizuka. Courtesy NASA.

the mission one more time and work through any questions. "We got to go over and spend an hour or two in the crew quarters with them. I spent most of my time with Mike Smith and Ellison Onizuka, who was my long-time friend from test pilot school. They were excited, and they were raunchy, as you would expect, and we had a lot of fun and a lot of good laughs. It was neat to go do that. So that was the last time that I got to physically go and sit with the crew and talk about the mission and the ascent and what to expect there."

On launch day the flight control team reported much earlier than the crew, monitoring the weather and getting ready for communication checks with the astronauts once they were strapped in. Covey said that he was excited to be working with Flight Director Jay Greene, whom he had worked with before, and that everything had seemed normal from his perspective leading into the launch. "From the control center standpoint," Covey said, "I don't remember anything that was unusual or extraordinary that we were working or talking about. It wasn't something where we knew that someone was making a decision and how they were making that decision. We just flat didn't have that insight. Didn't know what was going on. Did not. It was pretty much just everything's like a sim as we're sitting there getting ready to go."

Covey recalled that televisions had only recently been installed in the Mission Control Center and that the controllers weren't entirely sure what they were supposed to make of them yet. "The idea [had been] you shouldn't be looking at pictures; You should be looking at your data," he said. "So that's how we trained. Since the last time I'd been in the control center, they'd started putting [televisions in]. . . . I'd sat as the weather guy, and once the launch happens, I kind of look at the data, but I look over there at the TV."

Astronaut Fred Gregory was the weather CapCom for the 51L launch and recalled that nothing had seemed unusual leading up to the launch.

Up to liftoff, everything was normal. We had normal communication with the crew. We knew it was a little chilly, a little cold down there, but the ice team had gone out and surveyed and had not discovered anything that would have been a hazard to the vehicle. Liftoff was normal. . . . Behind the flight director was a monitor, and so I was watching the displays, but also every now and then look over and look at Jay Greene and then glance at the monitor. And I saw what appeared to be the solid rocket booster motor's explosive devices—what I thought—blew the solid rocket boosters away from the tank, and I was really surprised, because I'd never seen it with such resolution before, clarity before. Then I suddenly realized that what I was intellectualizing was something that would occur about a minute later, and I realized that a terrible thing had just happened.

Covey said Gregory's reaction was the first indication he had that something was wrong. "Fred is watching the video and sees the explosion, and he goes, 'Wha—? What was that?' Of course, I'm looking at my data, and the data freezes up pretty much. It just stopped. It was missing. So I look over and could not make heads or tails of what I was seeing, because I didn't see it from a shuttle to a fireball. All I saw was a fireball. I had no idea what I was looking at. And Fred said, 'It blew up,' something like that."

Covey recalled that the cameraman inside the control room continued to record what was happening there. "Amazingly, he's still sitting there just cranking along in the control center while this was happening. Didn't miss a beat," he said, "because I've seen too many film footages of me looking in disbelief at this television monitor trying to figure out what the hell it was I was seeing."

Off loop, Covey and Flight Director Jay Greene were talking, trying to gather information about what just happened. "There was a dialogue that started ensuing between Jay and myself," Covey recalled, *and Jay, he's trying to get confirmation on anything from anybody, if they have any data, and what they think has happened, what the status of the orbiter is. All we could get is the solid rocket boosters are separated. Don't know what else. I'm asking questions, because I want to tell the crew what to do. That's what the ascent CapComs are trained to do, is tell them what to do. If we know something that they don't, or we can figure it out faster, tell them so they can go and do whatever they need to do to recover or save themselves. There was not one piece of information that came forward; I was asking. I didn't do it over the loop, so I did this between Jay and some of the other people that could hear, "Are we in a contingency abort? If so, what type of contingency abort? Can we confirm they're off the* SRBs*?" Trying to see if there was anything I should say to the crew.*

In all the confusion, he said, no one said anything about him attempting to contact the crew members, since no one knew what to tell them. "We didn't have any comm. We knew that. That was pretty clear to me; so the only transmissions that I could have made would have been over a UHF [ultrahigh frequency], but if I didn't have anything to say to them, why call them? So we went through that for several minutes, and so if you go and look at it, there was never a transmission that I made after '*Challenger*, you're go [at] throttle up.' That was the last one, and there wasn't another one."

After a few minutes of trying to figure out if there was anything to tell the crew, reality started to hit. Covey said,

I remember Jay finally saying, "Okay, lock the doors. Everybody, no communications out. Lock the doors and go into our contingency modes of collecting data." I think when he did that, I finally realized; I went from being in this mode of, "What can we do? How do we figure out what we can do? What can we tell the crew? We've got to save them. We've got to help them save themselves. We've got to do something," to the realization that my friends had just died. . . . Of course, Fred and I were there together, which helped, because so many of the Challenger *crew were our classmates, and so we were sharing that together. A special time that I'll always remember being with Fred was there in the control center for that.*

It was a confusing time for those in the Launch Control Center. The data being received was not real-time data, Gregory said; there was a slight delay. "I had seen the accident occur on the monitor. I was watching data come in, but I saw the data then freeze, but I still heard the commentary about a normal flight coming from the public affairs person, who then, seconds later, stopped talking. So there was just kind of stunned silence in Mission Control."

"At this point," Gregory said, "no one had realized that we had lost the orbiter. Many, I'm sure, thought that this thing was still flying and that we had just lost radio signals with it. I think all of these things were kind of running through our minds in the first five to ten seconds, and then everybody realized what was going on."

What Covey and Gregory, relatively insular in their flight control duties, did not realize was that concerns over the launch had begun the day before. The launch had already been delayed six times, and because of the significance of the first Teacher in Space flight and other factors, many were particularly eager to see the mission take off. On the afternoon of 27 January (the nineteenth anniversary of the loss of astronauts Gus Grissom, Ed White, and Roger Chaffee in the *Apollo 1* pad fire), discussions began as to whether the launch should be delayed again. The launch complex at Kennedy Space Center was experiencing a cold spell atypical for the Florida coast, with temperatures on launch day expected to drop down into the low twenties Fahrenheit in the morning and still be near freezing at launch time.

During the night, discussions were held about two major implications of the cold temperatures. The first was heavy ice buildup on the launchpad and vehicle. The cold wind had combined with the supercooling of the cryogenic liquid oxygen and liquid hydrogen in the external tank to lead to the formation of ice. Concerns were raised that the ice could come off during flight and damage the vehicle, particularly the thermal protection tiles on the orbiter. A team was assigned the task of assessing ice at the launch complex.

The second potential implication was more complicated. No shuttle had ever launched in temperatures below fifty degrees Fahrenheit before, and there were concerns about how the subfreezing temperatures would affect the vehicle, and in particular, the O-rings in the solid rocket boosters. The boosters each consisted of four solid-fuel segments in addition to the nose cone with the parachute recovery system and the motor nozzle. The segments were assembled with rubberlike O-rings sealing the joints between

32. On the day of Space Shuttle *Challenger*'s 28 January 1986 launch, icicles draped the launch complex at the Kennedy Space Center. Courtesy NASA.

the segments. Each joint contained both a primary and a secondary O-ring for additional safety. Engineers were concerned that the cold temperatures would cause the O-rings to harden such that they would not fully seal the joints, allowing the hot gasses in the motor to erode them. Burn-through erosion had occurred on previous shuttle flights, and while it had never caused significant problems, engineers believed that there was potential for serious consequences.

During a teleconference held the afternoon before launch, SRB contractor Morton Thiokol expressed concerns to officials at Marshall Space Flight Center and Kennedy Space Center about the situation. During a second teleconference later that evening, Marshall Space Flight Center officials challenged a Thiokol recommendation that NASA not launch a shuttle at temperatures below fifty-three degrees Fahrenheit. After a half-hour off-line discussion, Thiokol reversed its recommendation and supported launch the next day. The three-hour teleconference ended after 11:00 p.m. (in Florida's eastern time zone).

During discussions the next morning, after the crew was already aboard the vehicle, orbiter contractor Rockwell International expressed concern that the ice on the orbiter could come off during engine ignition and ricochet and damage the vehicle. The objection was speculation, since no launch had taken place in those conditions, and the NASA Mission Management Team voted to proceed with the launch. The accident investigation board later reported that the Mission Management Team members were informed of the concerns in such a way that they did not fully understand the recommendation.

Launch took place at 11:38 a.m. on 28 January. The three main engines ignited seconds earlier, at 11:37:53, and the solid rocket motors ignited at 11:38:00. In video of the launch, smoke can be seen coming from one of the aft joints of the starboard solid rocket booster at ignition. The primary O-ring failed to seal properly, and hot gasses burned through both the primary and secondary O-rings shortly after ignition. However, residue from burned propellant temporarily sealed the joint. Three seconds later, there was no longer smoke visible near the joint.

Launch continued normally for the next half minute, but at thirty-seven seconds after solid rocket motor ignition, the orbiter passed through an area of high wind shear, the strongest series of wind shear events recorded thus far in the shuttle program. The worst of the wind shear was encountered at fifty-eight seconds into the launch, right as the vehicle was nearing "Max Q," the period of the highest launch pressures, when the combination of velocity and air resistance is at its maximum. Within a second, video captured a plume of flame coming from the starboard solid rocket booster in the joint where the smoke had been seen. It is believed that the wind shear broke the temporary seal, allowing the flame to escape. The plume rapidly became more intense, and internal pressure in the motor began dropping. The flame occurred in a location such that it quickly reached the external fuel tank.

The gas escaping from the solid rocket booster was at a temperature around six thousand degrees Fahrenheit, and it began burning through the exterior of the external tank and the strut attaching the solid rocket booster to the tank. At sixty-four seconds into the launch, the flame grew stronger, indicating that it had caused a leak in the external tank and was now burning liquid hydrogen escaping from the aft tank of the external tank. Approximately two seconds later, telemetry indicated decreasing pressure from the tank.

At this time, in the vehicle and in Mission Control, the launch still appeared to be proceeding normally. Having made it through Max Q, the vehicle throttled its engines back up. At sixty-eight seconds, Covey informed the crew it was "Go at throttle up." Commander Dick Scobee responded, "Roger, go at throttle up," the last communication from the vehicle.

Two seconds later, the flame had burned through the attachment strut connecting the starboard SRB and the external tank. The upper end of the booster was swinging on its strut and impacted with the external tank, rupturing the liquid oxygen tank at the top of the external tank. An orange fireball appeared as the oxygen began leaking. At seventy-three seconds into the launch, the crew cabin recorder records Pilot Michael Smith on the intercom saying, "Uh-oh," the last voice recording from *Challenger*.

While the fireball caused many to believe that the Space Shuttle had exploded, such was not the case. The rupture caused the external tank to lose structural integrity, and at the high velocity and pressure it was experiencing, it quickly began disintegrating. The two solid rocket boosters, still firing, disconnected from the shuttle stack and flew freely for another thirty-seven seconds. The orbiter, also now disconnected and knocked out of proper orientation by the disintegration of the external tank, began to be torn apart by the aerodynamic pressures. The orbiter rapidly broke apart over the ocean, with the crew cabin, one of the most solid parts of the vehicle, remaining largely intact until it made contact with the water.

All of that would eventually be revealed during the course of the accident investigation. At Mission Control, by the time the doors were opened again, much was still unknown, according to Covey.

[We] had no idea what had happened, other than this big explosion. We didn't know if it was an SRB that exploded. I mean, that was what we thought. We always thought SRBs would explode like that, not a big fireball from the external tank propellants coming together. So then that set off a period then of just trying to deal with that and the fact that we had a whole bunch of spouses and families that had lost loved ones and trying to figure out how to deal with that.

The families were in Florida, and I remember, of course, the first thing I wanted to do was go spend a little time with my family, and we did that. But then we knew the families were coming back from Florida and out to Ellington [Field, Houston], so a lot of us went out there to just be there when they came back in. I remember it was raining. Generally they were keeping them isolated, but a big crowd of us waiting for them, they loaded them up to come home. Then over the next several days most of the time we spent was trying to help the Onizukas in some way; being around. Helping them with their family as the families flew in and stuff like that.

After being in Mission Control for approximately twelve hours—half of that prior to launch and the rest in lockdown afterward, analyzing data, Gregory finally headed home.

The families had all been down at the Kennedy Space Center for the liftoff, and they were coming back home. Dick Scobee, who was the commander, lived within a door or two of me. And when I got home, I actually preceded the families getting home; I remember that. They had the television remote facilities already set up outside of the Scobees' house, and it was disturbing to me, and so I went over and, in fact, invited some of those [reporters] over to my house, and I just talked about absolutely nothing to get them away from the house, so that when June Scobee and the kids got back to the house, they wouldn't have to go through this gauntlet.

The next few days, Gregory said, were spent protecting the crew's families from prying eyes. "There was such a mess over there that Barbara and I took [Scobee's] parents and just moved them into our house, and they must have stayed there for about four or five days. Then June Scobee, in fact, came over and stayed, and during that time is when she developed this concept for the Challenger Center. She always gives me credit for being the one who encouraged her to pursue it, but that's not true. She was going to do it, and it was the right thing to do."

Gregory recalled spending time with the Scobees and the Onizukas and the Smiths, particularly Mike Smith's children. "It was a tough time," he said.

It was a horrible time, because I had spent a lot of time with Christa McAuliffe and [her backup] Barbara Morgan, and the reason was because I had teachers in my family. On my father's side, about four or five generations; on my mother's side, a couple of generations. My mother was elementary school, and my dad was more in the high school. But Christa and I and Barbara talked about how important it was, what she was doing, and then what she was going to do on orbit and how it would be translated down to the kids, but then what she was going to do once she returned. So it was traumatic for me, because not only had I lost these longtime friends, with Judy Resnik and Onizuka and Ron McNair and Scobee, and then Mike Smith, who was a class behind us, but I had lost this link to education when we lost Christa.

Astronaut Sally Ride was on a commercial airliner, flying back to Houston, when the launch tragedy occurred. "It was the first launch that I hadn't seen, either from inside the shuttle or from the Cape or live on television," Ride recalled.

The pilot of the airline, who did not know that I was on the flight, made an announcement to the passengers, saying that there had been an accident on the Challenger. *At the time, nobody knew whether the crew was okay; nobody knew what had happened. Thinking back on it, it's unbelievable that the pilot made the announcement he made. It shows how profoundly the accident struck people. As soon as I heard, I pulled out my NASA badge and went up into the cockpit. They let me put on an extra pair of headsets to monitor the radio traffic to find out what had happened. We were only about a half hour outside of Houston; when we landed, I headed straight back to the Astronaut Office at JSC.*

Payload Specialist Charlie Walker was returning home from a trip to San Diego, California, when the accident occurred.

I can remember having my bags packed and having the television on and searching for the station that was carrying the launch. As I remember it, all the stations had the launch on; it was the Teacher in Space mission. So I watched the launch, and to this day, and even back then I was still aggravated with news services that would cover a launch up until about thirty seconds, forty-five sec-

onds, maybe one minute in flight, after Max Q, and then most of them would just cut the coverage. "Well, the launch has been successful." [I would think,] "You don't know what you're talking about. You're only thirty seconds into this thing, and the roughest part is yet to happen."

And whatever network I was watching ended their coverage. "Well, looks like we've had a successful launch of the first teacher in space." And they go off to the programming, and it wasn't but what, ten seconds later, and I'm about to pick my bags up and just about to turn off the television and go out my room door when I hear, "We interrupt this program again to bring you this announcement. It looks like something has happened." I can remember seeing the long-range tracker cameras following debris falling into the ocean, and I can remember going to my knees at that point and saying some prayers for the crew. Because I can remember the news reporter saying, "Well, we don't know what has happened at this point." I thought, "Well, you don't know what has happened in detail, but anybody that knows anything about it can tell that it was not at all good."

Mike Mullane was undergoing payload training with the rest of the 62A crew at Los Alamos Labs in New Mexico. "We were in a facility that didn't have easy access to a TV," said Mullane.

We knew they were launching, and we wanted to watch it, and somebody finally got a television or we finally got to a room and they were able to finagle a way to get the television to work, and we watched the launch, and they dropped it away within probably thirty seconds of the launch, and we then started to turn back to our training. Somebody said, "Well, let's see if they're covering it further on one of the other channels," and started flipping channels, and then flipped it to a channel and there was the explosion, and we knew right then that the crew was lost and that something terrible had happened.

Mullane theorized that someone must have inadvertently activated the vehicle destruction system or a malfunction caused the flight termination system to go off. "I was certain of it," Mullane said.

I mean, the rocket was flying perfectly, and then it just blew up. It just looked like it had been blown up from this dynamite. Shows how poor you can be as a witness to something like this, because that had absolutely nothing to do with it.

But it was terrible. Judy was killed on it. She was a close friend. There were four people from our group that were killed. It was a terrible time. Really as bad as it gets. It was like a scab or a wound that just never had an opportunity to heal because you had that trauma.

Astronaut Mary Cleave recalled two very different sets of reactions to the tragedy from the people around her in the Astronaut Office. "For the guys in the corps, when you're in the test pilot business, you're sort of a tough guy," she said.

It's a part of the job. It's a lousy part of the job, but it's part of the job. But I mean, the secretaries and everybody else were really upset, so we spent some time with them. Before my first flight, I had signed up. I basically told my family, "Hey, I might not be coming back." When we flew, it was the heaviest payload to orbit. We were already having nozzle problems. I think a lot of us understood that the system was really getting pushed, but that's what we'd signed up to do. I think probably a lot of people in the corps weren't as surprised as a lot of other people were. I did crew family escort afterwards. I was assigned to help when the families came down, as an escort at JSC *when the president came in to do the memorial service. Jim Buchli was in charge of the group; they put a marine in charge of the honor guard. So I got to learn to be an escort from a marine, which was interesting. I learned how to open up doors. This was sort of like it doesn't matter if you're a girl or boy, there's a certain way people need to be treated when they're escorted. So I did that. That was interesting. And it was nice to think that you could help at that point.*

Charlie Bolden had just returned to Earth ten days earlier from his first spaceflight, 61C. His crew was wrapping up postmission debriefing, he recalled, and it gathered with others in the Astronaut Office to watch *Challenger* launch. "That was the end of my first flight, and we were in heaven. We were celebrating as much as anybody could celebrate," he said. "We sat in the Astronaut Office, in the conference room with everybody else, to watch *Challenger*. Nobody was comfortable because of all the ice on the launchpad and everything. I don't think there were many of us who felt we should be flying that day, but what the heck. Everybody said, 'Let's go fly.' And so we went and flew."

Bolden thought the explosion was a premature separation of the solid rocket boosters; he expected to see the vehicle fly out of the smoke and perform a return-to-launch-site abort. "We were looking for something good to come out of this, and nothing came out except these two solid rocket boosters going their own way."

It took awhile, but it finally sunk in: the vehicle and the crew were lost. "We were just all stunned, just didn't know what to do," Bolden said. "By the end of the day we knew what had happened; we knew what had caused the accident. We didn't know the details, but the launch photography showed us the puff of smoke coming out of the joint on the right-hand solid rocket booster. And the fact that they had argued about this the night before meant that there were people from [Morton] Thiokol who could say, 'Let me tell you what happened. This is what we predicted would happen.'"

Bolden was the family escort for the family of 51L mission specialist Ronald McNair. Family escorts are chosen by crew members to be with the families during launch activities and to be a support to families if something happened to the crew, as was the case with 51L. Much of Bolden's time in the year after the incident was spent helping the McNair family, which included Ron's wife and two children. "I sort of became a surrogate, if you will, for [McNair's children] Joy and Reggie, and just trying to make sure that Cheryl [McNair] had whatever she needed and got places when she was supposed to be there. Because for them it was an interminable amount of time, I mean years, that they went through the postflight grieving process and memorial services and that kind of stuff."

Bolden's 61C crewmate Pinky Nelson was on his way to Minneapolis, Minnesota, for the premier of the IMAX movie *The Dream Is Alive*, which included footage from Nelson's earlier mission, 41C. Nelson recalled having worked closely with the 51L crew, which Nelson said would be using the same "rinky-dink little camera" as his crew to observe Halley's Comet.

I'd spent a bunch of time trying to teach Ellison [Onizuka] how to find Halley's Comet in the sky. [Dick] Scobee and I were really close friends because of 41C, so "Scobe" and I had talked a lot about his kind of a "zoo crew," about his crew and all their trials and tribulations. He really wanted to get this mission flown and over with. So I talked to them the night before, actually, from down at the Cape and wished them good luck and all that, and then the accident happened while I was on the airplane to Minneapolis.

Nelson flew back to Houston from Minneapolis that afternoon, arriving around the same time that the families were arriving from Florida. Nelson and his wife, Susie, and astronaut Ox van Hoften and his wife convened at the Scobees' home.

"The national press was just god-awful," Nelson said.

I've never forgiven some of those folks. . . . I mean, it's their job, but still—for their just callous, nasty behavior. We just spent a lot of time just kind of over at Scobee's, trying to just be there and help out. I still can't drink flavored coffee. That's the only kind of coffee June had, vanilla bean brew or something. So whenever I smell that stuff, that's always my memory of that, is having bad coffee at Scobe's house, trying to just get their family through the time, just making time pass. We had to unplug the phones. The press was parked out in front of the house. It was a pretty bad time for all that. We went over and tried to do what we could with some of the other families. My kids had been good friends with Onizuka's kids; they're the same age. Lorna [Onizuka] was having just a really hard time. Everyone was trying to help out where we could.

Memorial services were beginning to be held for the lost crew members even as the agency was continuing with its search for the cockpit and the bodies of the lost crew. "It was terrible, going to the memorial services," admitted Mullane.

It was one of those things that didn't seem to end, because then they were looking for the cockpit out there. I personally thought, "Why are we doing this? Leave the cockpit down there. What are you going to learn from it?" Because by then they knew the SRB *was the problem. . . . I remember thinking, "Why are we even looking for that cockpit? Just bury them at sea. Leave them there." I'm glad they did, though, because later I heard it was really shallow where that cockpit was. It was like, I don't know, like eighty feet or something, which is too shallow, because somebody eventually would have found it and pulled it up on a net or been diving on it or something. So it's good that they did look for it. So you had these several weeks there, and then they bring that cockpit up, and then you have to repeat all the memorial services again, because now you have remains to bury. And then plus on top of that, you had the revelation that it wasn't an accident; it was a colossal screwup. And you had that to deal with. So it was a miserable time, about as bad as I've ever lived in my life, were those months surrounding, months and years, really, surrounding the* Challenger *tragedy.*

Astronaut Bryan O'Connor was at Kennedy Space Center during the debris recovery efforts and postrecovery analysis. O'Connor recalled being on the pier when representatives from the Range Safety Office at Cape Canaveral were trying to determine whether what happened was an inadvertent range safety destruct—if somehow there had been a malfunction of the destruct package intended to destroy the vehicle should a problem cause it to pose a safety risk to those on the ground. "I remember there was a Coast Guard cutter that came in and had some pieces and parts of the external tank," O'Connor said. "On the second or third day, I think, one of these ships actually had a piece of the range safety destruct system from the external tank, intact for about halfway and then ripped up the other half of it. When he looked at that, he could tell that it hadn't been a destruct."

O'Connor had accident investigation training and was then assigned to work with Kennedy Space Center on setting up a place to reconstruct the vehicle as debris was recovered. "I remember we put tape down on the floor. We got a big room in the Logistics Center. They moved stuff out of the way. As time went on, the need increased for space, and we actually ended up putting some things outside the Logistics Center, like the main engines and some of the other things. But the orbiter pretty much was reassembled piece by piece over a period of time as the parts and pieces were salvaged out of the water, most of them floating debris, but some, I think, was picked up from subsurface."

Recovery efforts started with just a few ships, O'Connor said, but grew into a large fleet. According to the official Rogers Commission Report on the accident, sixteen watercraft assisted in the recovery, including boats, submarines, and underwater robotic vehicles from NASA, the navy, and the air force. "It was one of the biggest salvage efforts ever, is what I heard at the time," O'Connor said. "Over a period of time, we were able to rebuild quite a bit of the orbiter, laying it out on the floor and, in some cases, actually putting it in a vertical structure. Like the forward fuselage, for example, we tried to make a three-dimensional model from the pieces that we recovered there."

While the goal had originally been to determine the cause of the accident, the investigation eventually shifted to its effects, with analysis of the

33. This photograph, taken a few seconds after the loss of *Challenger*, shows the Space Shuttle's main engines and solid rocket booster exhaust plumes entwined around a ball of gas from the external tank. Courtesy NASA.

debris revealing how the various parts of the vehicle had been affected by the pressures during its disintegration.

Astronaut Joe Kerwin, a medical doctor before his selection to the corps and a member of the first crew of the Skylab space station, was the director of Space Life Sciences at Johnson Space Center at the time of the accident. "Like everybody else at JSC I remember exactly where I was when it happened," Kerwin recalled.

I didn't see it live. In fact, I was having a staff meeting in my office at JSC and we had a monitor in the background because the launch was taking place. And I just remember all of us sort of looking up and seeing this explosion taking place on the monitor. And there was the moment of silence as each of us tried to absorb what it looked like was or might be going on, and then sort of saying, "Okay, guys, I think we better get to work. We're going to need to coordinate with the Astronaut Office. We're going to have to have flight surgeons. Sam, you contact the doctors on duty down at the

Cape and make sure that they have the families covered," and we just sort of set off like that.

His medical team's first actions were simply to take care of the families of the crew members, Kerwin said. "Then as the days went by and the search for the parts of the orbiter was underway I went down to Florida and coordinated a plan for receiving bodies and doing autopsies and things of that nature. It included getting the Armed Forces Institute of Pathology to commit to send a couple of experts down if and when we found remains to see whether they could determine the cause of death."

But the crew compartment wasn't found immediately and Kerwin went back home to Houston, where days turned into weeks.

I was beginning to almost hope that we wouldn't have to go through that excruciating investigation when I had a call from Bob Crippen that said, no, we've found the crew compartment and even at this late date there are going to be some remains so how about let's get down here. I went down immediately.

By that time the public and press response to the accident and to NASA had turned bad, and NASA, which had always been considered one of the best organizations in government, was now one of the worst organizations in government and there was a lot of bad press and there were a lot of paparazzi there in Florida who just wanted to get in on the action and get gruesome pictures or details or whatever they could. So we had to face that.

In addition to dealing with the press, Kerwin said recovery efforts also had to deal with local politics.

The local coroner was making noises like this accident had occurred in his jurisdiction and therefore he wanted to take charge of any remains and perform the autopsies himself, which would have been a complicating factor, to say the least. I didn't have to deal with that, I just knew about it and that I might have had to deal with [the] coroner if the offensive line didn't block him. But the higher officials in NASA and in particular in the State of Florida got him called off, saying, "No, this accident was in a federal spacecraft and it occurred offshore and you just back off."

By the time Kerwin arrived at the scene, the recovery team had come up with several different possible ways to get the remains to where they would be autopsied.

In view of the lateness of the time and in view of the press coverage and all that stuff, we decided that we needed a much more secure location for this activity to take place and we were given space in one of the hangars at Cape Canaveral, one of the hangars in which the Mercury crews trained way back in the early sixties. We quickly prepared that space for the conduct of autopsies. When the remains were brought in they were brought into one of the piers by motorized boat. It was done after dark, and there was always one or two or three astronauts with the remains and we put up a little screen because right across the sound from this pier the press had set up floodlights and cameras and bleachers.

An unmarked vehicle was used to transport the remains to the hangar, while a NASA ambulance was used as a decoy to keep the location of the autopsies a secret. "We knew by that time, and really knew from the beginning, that crew actions or lack of actions didn't have anything to do with the cause of the *Challenger* accident," Kerwin said.

As an accident investigator you ask yourself that first—could this have been pilot error involved in any way, and the immediate answer was no. But we still needed to do our best to determine the cause of death, partly because of public interest but largely because of family interest in knowing when and under what circumstances their loved ones died. So that was the focus of the investigation.

The sort of fragmentary remains that were brought in, having been in the water for about six weeks, gave us no clue as to whether the cause of death was ocean impact or whether it could've taken place earlier. So the only thing left for the autopsies was to determine, was to prove, that each of the crew members had had remains recovered that could identify that, yes, that crew member died in that accident. That was not easy but we were able to do it. So at the end of the whole thing I was able to send a confidential letter to the next of kin of each of the crew members stating that and stating what body parts had been recovered, a very, very short letter.

In addition to identifying the remains, Kerwin and his team worked to identify the exact cause of death of the crew members.

We'd all seen the breakup on television, we knew it was catastrophic, and my first impression as a doctor was that it probably killed all the crew just right at the time of the explosion. But as soon as they began to analyze the camera foot-

age, plus what very little telemetry they had, it became apparent that the g-forces were not that high, not as high as you'd think. I guess the crew compartment was first flung upward and away from the exploding external tank and then rapidly decelerated by atmospheric pressure until it reached free fall. The explosion took place somewhere about forty thousand feet, I think about forty-six or forty-seven thousand, and the forces at breakup were estimated to be between fifteen and twenty gs, which is survivable, particularly if the crew is strapped in properly and so forth. The crew compartment then was in free fall, but upward. It peaked at about sixty-five thousand feet, and about two and half minutes after breakup hit the ocean at a very high rate of speed. If the crew hadn't died before, they certainly died then.

Kerwin said he and his team then worked to refine their understanding of exactly what had happened when. "Our investigation attempted to determine whether or not the crew compartment's pressure integrity was breeched by the separation, and if so, if the crew compartment had lost pressure, we could then postulate that the crew had become profoundly unconscious because the time above forty thousand feet was long enough and the portable emergency airpack was just that, it was an airpack, not an oxygen pack, so they had no oxygen available. They would have become unconscious in a matter of thirty seconds or less depending on the rate of the depressurization and they would have remained unconscious at impact."

But the damage to the compartment was too great to allow Kerwin to determine with certainty whether the cabin had lost pressure. He said that, given what it would mean for the crew's awareness of its fate, his team was almost hopeful of finding evidence that the crew compartment had been breached, but they were unable to make a conclusive determination.

The damage done to the crew compartment at water impact was so great that despite a really lot of effort, most of it pretty expert, to determine whether there was a pressure broach in any of the walls, in any of the feed-throughs, any of the windows, any of the weak points where you might expect it, we couldn't rule it out but we couldn't demonstrate it either. We lifted all the equipment to see whether any of the stuff in the crew compartment looked as if it had been damaged by rapid decompression, looked at toothpaste tubes and things like that, tested some of them or similar items in vacuum chambers to determine wheth-

er that sort of damage was pressure-caused and that was another blind hole. We simply could not determine. And then as to the remains, in a case of immediate return of the remains we might have gotten lucky and been able to measure tissue oxygen concentration and tissue carbon dioxide concentration and gotten a clue from that but this was way too late for that kind of thing.

Finally, after all of that we had to just, I just had to sit down and write the letter to Admiral [Richard H.] Truly stating that we did not know the cause of death.

Sally Ride was named to the commission that was appointed to investigate the causes of the accident. President Reagan appointed former secretary of state William Rogers to head a board to determine the factors that had resulted in the loss of *Challenger*. (Ride has the distinction of being the only person to serve on the presidential commission investigations of both the *Challenger* and the *Columbia* accidents.) Ride was in training at the time of the accident and was assigned to the presidential commission within just a few days of the accident. The investigation lasted six months.

"The panel, by and large, functioned as a unit," Ride recalled.

We held hearings; we jointly decided what we should look into, what witnesses should be called before the panel, and where the hearings should be held. We had a large staff so that we could do our own investigative work and conduct our own interviews. The commission worked extensively with the staff throughout the investigation. There was also a large apparatus put in place at NASA to help with the investigation: to analyze data, to look at telemetry, to look through the photographic record, and sift through several years of engineering records. There was a lot of work being done at NASA under our direction that was then brought forward to the panel. I participated in all of that. I also chaired a subcommittee on operations that looked into some of the other aspects of the shuttle flights, like was the astronaut training adequate? But most of our time was spent on uncovering the root cause of the accident and the associated organizational and cultural factors that contributed to the accident.

Ride said she had planned to leave NASA after her upcoming third flight and do research in an academic setting. But her role in the accident investigation caused her to change her plans. "I decided to stay at NASA for an extra year, simply because it was a bad time to leave," explained Ride.

She described that additional year as one that was very difficult both for her personally and for the corps. "It was a difficult time for me and a difficult time for all the other astronauts, for all the reasons that you might expect," said Ride. "I didn't really think about it at the time. I was just going from day to day and just grinding through all the data that we had to grind through. It was very, very hard on all of us. You could see it in our faces in the months that followed the accident. Because I was on the commission, I was on TV relatively frequently. They televised our hearings and our visits to the NASA centers. I looked tired and just kind of gray in the face throughout the months following the accident."

Astronaut Steve Hawley, who was married to Ride at the time, recalled providing support to the Rogers Commission, particularly in putting together the commission's report, *Leadership and America's Future in Space.*

The chairman felt that it was appropriate to look not only at the specifics of that accident but other things that his group might want to say about safety in the program, and that included, among other things, the role of astronauts in the program, and that was one of the places where I think I contributed, was how astronauts ought to be involved in the program. I remember one of the recommendations of the committee talked about elevating the position of director of FCOD [Flight Crew Operations Directorate], because at the time of the Challenger *accident, he was not a direct report to the center director. That had been a change that had been made sometime earlier, I don't remember exactly when. And that commission felt that the guy that was head of the organization with the astronauts should be a direct report to the center director. So several people went back to Houston and put George [Abbey]'s desk up on blocks in an attempt to elevate his position. I think he left it there for some period of time, as far as I can remember.*

Though not everyone in the Astronaut Office was involved in the Rogers Commission investigation and report, many astronauts were involved in looking into potential problems in the shuttle hardware and program. Recalled O'Connor,

I remember that a few days after the accident Dick Truly and the acting administrator [William Graham] were down at KSC. Again, because I had accident investigation training, they had a discussion about what's the next steps

here. The acting administrator, it turned out, was very reluctant to put a board together, a formal board. We had an Accident Investigation Team that was assembling; eventually became under the cognizance of J. R. Thompson. We had a bunch of subteams under him, and then he was reporting to Dick Truly. But we never called out a formal board, and I think there was more or less a political feel that this is such high visibility that we know for sure that the Congress or the president or both are going to want to have some sort of independent investigation here, so let's not make it look like we're trying to investigate our own mishap here. And that's why he decided not to do what our policy said, which is to create a board. We stopped short of that.

Once the presidential commission was in place, O'Connor said it was obvious to Dick Truly that he needed to provide a good interface between the commission and NASA. Truly assigned O'Connor to set up an "action center" in Washington DC for that purpose. The center kept track of all the requests from the commission and the status of the requests.

"They created a room for me, cleared out all the desks and so on," O'Connor recalled. "We put [up] a bunch of status boards; very old-fashioned by today's standard, when I think about it. It was more like World War II's technology. We had chalkboards. We had a paper tracking system, an IBM typewriter in there, and so on. It all seemed so ancient by today's standards, virtually nothing electronic. But it was a tracking system for all the requests that the board had. It was a place where people could come and see what the status of the investigation was."

The center became data central for NASA's role in the investigation. In addition to the data keeping done there, Thompson would have daily teleconferences with all of his team members to find out what was going on.

Everybody would report in, what they had done, where they were, where they were on the fault tree analysis that we were doing to x out various potential cause factors. I remember the action center became more than just a place where we coordinated between NASA and the blue-ribbon panel. It was also a place where people could come from the [Capitol] Hill or the White House. We had quite a few visitors that came, and Dick Truly would bring them down to the action center to show them where we were in the investigation. So it had to kind of take on that role, too, of publicly accessible communication device.

I did that job for some weeks, and then we rotated people. [Truly] asked that George Abbey continue [to provide astronauts]. George had people available now that we're not going to fly anytime soon, so he offered a bunch of high-quality people. . . . So shortly after I was relieved from that, I basically got out of the investigation role and into the "what are we going to do about it" role and was assigned by George to the Shuttle Program Office.

As board results came out, changes began to be made. For example, O'Connor said, all of the mission manifests from before the accident were officially scrapped.

I was relieved of my job on a crew right away. I think they called it 61M, which was a mission I was assigned to right after I got back from 61B. I can't remember who all the members were, but my office mate, Sally Ride, and I were both assigned to that same mission, 61M. Of course, when the accident happened, all that stuff became questionable and we stopped training altogether. I don't think that mission ever resurrected. It may have with some other name or number, but the crew was totally redone later, and the two of us got other assignments.

In the meantime, O'Connor was appointed to serve as the assistant program manager for operations and safety. His job included coordinating how NASA was going to respond to a couple of the major recommendations that came out of the blue-ribbon panel. The panel had ten recommendations, covering a variety of subjects. One of them had to do with how to restructure and organize the safety program at NASA. Another one O'Connor was involved with dealt with wheels, tires, brakes, and nose-wheel steering.

"It was all the landing systems," he explained.

Now, that may sound strange, because that had nothing to do with this accident, but the Challenger *blue-ribbon panel saw that, as they were looking at our history on shuttle, they saw that one of the bigger problems we were addressing technically with that vehicle was landing rollout. We had a series of cases where we had broken up the brakes on rollout by overheating them or overstressing them. We had some concerns about automatic landings. We had some concerns about steering on the runway in cases of a blown tire or something like that. So they chose to recommend that we do something about these things, put more emphasis on it, make some changes and upgrades in that area.*

In the wake of the accident, even relatively low-profile aspects of the shuttle program were reexamined and reevaluated. "NASA wanted to go back and look at everything, not just the solid rocket boosters, but everything, to determine, is there another *Challenger* awaiting us in some other system," recalled Mullane, who was assigned to review the range safety flight termination system, which would destroy the vehicle in the event it endangered lives. "I always felt it was a moral issue on this dynamite system. I always felt it was necessary to have that on there, because your wives and your family, your LCC [Launch Control Center] people are sitting there two and a half miles away. If you die, that's one thing. But if in the process of you dying that rocket lands on the LCC and kills a couple of hundred people, that's not right. So they should have dynamite aboard it to blow it up in case it is threatening the civilian population."

While Mullane himself supported having the flight termination system on the vehicles, he said that some of his superiors did not.

I could not go to the meetings and present their position. I couldn't. I mean, to me it was immoral. It was immoral to sit there and say we fly without a dynamite system aboard. That's immoral. And we threatened LCC, we threaten our families, we threaten other people. We're signing up for the risk to ride in the rocket.

That was another bad time of my life, because I took a position that was counter to my superiors' position, and I felt that it was jeopardizing my future at NASA. I didn't like that at all, didn't like the idea that I was supposed to just parrot somebody's opinion and mine didn't count on that issue. . . . So I did not like that time of my life at all. I had an astronaut come to me once . . . who heard my name being kicked around as a person that was causing some problems. And that's the last thing you want in your career is to hear that your name is in front of people who make launch crew decisions, who make crew decisions, and basically, it looks like I'm a bad apple. But I just couldn't do it. . . . I said, "You go. You do it. I can't. It's immoral." By the way, the end of that is that the solid boosters retained their dynamite system aboard. It was taken off the gas tank, making it much safer, at least now two minutes up when the boosters are gone, we don't have to dynamite aboard anymore, so it could fail and blow you up. But that's the right decision right there. You protect the civilians, you protect your family, you protect LCC with that system.

The results from the commission questioned the NASA culture, and according to Covey some within the agency found that hard to deal with, especially while still recovering from the loss of the crew and vehicle. "Everybody was reacting basically to two things," Covey said. "One was the fact that they had lost a Space Shuttle and lost a crew, and two, the Rogers Commission was extremely critical, and in many cases, rightfully so, about the way the decision-making processes had evolved and the culture had evolved."

Covey noted,

Those two things together are hard for any institution to accept, because this was still a largely predominant workforce that had come through the Apollo era into the shuttle era and had been immensely successful in dealing with the issues that had come through both those programs to that point. So to be told that the culture was broken was hard to deal with, and that's because culture doesn't change overnight, and there was a lot of people that didn't believe that that was an accurate depiction of the situation and environment that existed within the agency, particularly at the Johnson Space Center.

Personnel changes began to take place, including the departures of George Abbey, the long-time director of flight crew operations, and John Young, chief of the Astronaut Office. "We started seeing a lot of personnel changes in that time period in leadership positions," Covey said.

I think they were a matter of timing and other things, but George Abbey had been the director of flight crew operations for a long time, and somewhere in there George left that position. Don Puddy came in as the director of flight crew operations, and that sat poorly with a lot of people, because he wasn't out of Flight Crew Operations; he was a flight director. So that was something that a lot of people just had a hard time accepting. John Young left from being chief of the Astronaut Office, and Dan Brandenstein came into that position. I'm trying to remember what happened in Mission Operations, but somewhere in there Gene Kranz left; I can't remember when, and I'm not sure when in that spectrum of things that he did. So there were changes there. The center director changed immediately after the Challenger *accident, also, and changes were rampant at headquarters. So, basically, there was a restructuring of the leadership team from the administrator on down.*

In time, the astronaut corps grieved and adjusted to leadership changes, and their focus returned to flying. "As JSC does," Covey said, "once they got past the grief and got past the disappointment of failure and acceptance of their role relative to the decision-making process, they jumped in and said, 'Okay, our job is flying. Let's go figure out how we're going to fly again.'"

Covey said the *Challenger* accident caused some in the astronaut corps to be wary of the NASA leadership structure and to distrust that the system would make the right kinds of decisions to protect them. "I wouldn't say it was a bunker mentality, but it was close to that," Covey said.

The idea that, as it played out, that there were decisions made and information that may not have been fully considered, and as you can see from all of it, a relatively limited involvement of any astronauts or flight crew people in the decision that led up to the launch; very little, if any. So that led to some changes that have evolved over time where there are more and more astronauts that have been involved in that decision-making process at the highest levels, either within the Space Shuttle program or in the related activities, where before it was like, "Yeah, those guys will make the right decision, and we'll go fly."

Along those same lines, Fred Gregory commented that the tragedy was a moment of realization for many about the dangers of spaceflight.

The first four missions we called test flights, and then on the fifth mission we declared ourselves operational. We were thinking of flying journalists, and we had Pete Aldridge, who was the secretary of the air force. [about to fly]. I mean, we were thinking of ourselves as almost like an airline at that point. It came back safely. Everything was okay, even though there may have been multiple failures or things that had degraded. When we looked back, we saw that, in fact, we had had this erosion of those primary and secondary O-rings, but since it was a successful mission and we came back, it was dismissed almost summarily. I think there was a realization that we were vulnerable, and that this was not an airliner, flying to space was risky, and that we were going to have to change the approach that we had taken in the past.

During the investigation, new light was shined on past safety issues and close calls. Astronaut Don Lind recalled being informed of a situation that occurred on his 51B mission, less than a year earlier.

What happened was that Bob Overmyer and I had shared an office for three and a half years, getting ready for this mission. Bob, when they first started the

Challenger *investigation, was the senior astronaut on the investigation team. He came back from the Cape one day, walked in the office, slumped down in the chair, and said, "Don, shut the door." Now, in the Astronaut Office, if you shut the door, it's a big deal, because we tried to keep an open office so people could wander in and share ideas. . . . So I shut the door, and Bob said, "The board today found out that on our flight nine months previously we almost had the same explosion. We had the same problems with the O-rings on one of our boosters." We talked about that for a few minutes, and he said, "The board thinks we came within fifteen seconds of an explosion."*

After learning this, Lind traveled out to what was then Morton Thiokol's solid rocket booster facility in Utah to find out exactly what had happened on his flight.

Alan McDonald, who was the head of the Booster Division, sat down with me. . . . He got out his original briefing notes for Congress, which I now have, and outlined exactly what had happened. There are three separate O-rings to seal the big long tubes with the gases flowing down through them at about five thousand degrees and 120 psi. The first two seals on our flight had been totally destroyed, and the third seal had 24 percent of its diameter burned away. McDonald said, "All of that destruction happened in six hundred milliseconds, and what was left of that last O-ring, if it had not sealed the crack and stopped that outflow of gases, if it had not done that in the next two hundred to three hundred milliseconds, it would have been gone all the way. You'd never have stopped it, and you'd have exploded. So you didn't come within fifteen seconds of dying, you came within three-tenths of one second of dying." That was thought provoking.

In the wake of the *Challenger* disaster, all future missions were scrubbed and the flight manifest was replanned. Some missions were rescheduled with new numbers and only minor changes, while other planned missions and payloads were canceled entirely.

Recalled Charlie Walker of the electrophoresis research he had been conducting, "The opportunity just went away with the national policy changes; commercial was fourth priority, if it was a priority at all, for shuttle manifest. . . . Shuttle would not be flying with the regularity or the frequency that had been expected before."

The plans to use the Centaur upper stage to launch the *Ulysses* and *Galileo* spacecraft from the shuttle were also canceled as a result of the accident. Astronaut Mike Lounge had been assigned to the *Ulysses* Centaur flight, which was to have flown in spring 1986 on *Challenger*. "So when we saw *Challenger* explode on January 28, before that lifted off, I remember thinking, 'Well, Scobee, take care of that spaceship, because we need it in a couple of months.' We would have been on the next flight of *Challenger*."

Lounge recalled that his crew was involved in planning for the mission during the *Challenger* launch. They took a break from discussing the risky Centaur mission, including ways to eject the booster if necessary, so that they could watch the 51L launch on a monitor in the meeting room. Lounge said the disaster shone a new light on the discussions they'd been having. "We assumed we could solve all these problems. We were still basically bulletproof. Until *Challenger*, we just thought we were bulletproof and the things would always work."

Rick Hauck, who was assigned to command the *Ulysses* Centaur mission, was glad to see the mission canceled. "Would it have gotten to the point where I would have stood up and said, 'This is too unsafe. I'm not going to do it'?" Hauck said. "I don't know. But we were certainly approaching levels of risk that I had not seen before."

For Bob Crippen, the *Challenger* accident would result in another loss, that of his final opportunity for a shuttle flight. The ramifications of the accident meant that he ended his career as a flight-status astronaut the same way he began it—in the crew quarters for Vandenberg's SLC-6 launch complex, waiting for a launch that would never come. In the 1960s that launch had been of the Air Force's Manned Orbiting Laboratory. In 1986 it was shuttle mission 62A.

"If I have one flying regret in my life, it was that I never had an opportunity to do that Vandenberg mission," Crippen said.

We [had been] going to use filament-wound solid rockets, whereas the solid rockets that we fly on board the shuttle have steel cases. That was one of the things, I think, that made a lot of people nervous. . . . We needed the filament-wounds to get the performance, the thrust-to-weight ratio that we needed flying out of Vandenberg. So they used the filament-wound to take the weight out. After we had the joint problem on the solids with Challenger*; most people just couldn't* *get comfortable with the filament-wound case, so that was one of the aspects of why they ended up canceling it.*

As the agency, along with the nation, was reevaluating what the future of human spaceflight would look like, what things should be continued, and what things should be canceled, the astronauts in the corps had to make the same sorts of decisions themselves. Many decided that the years-long gap in flights offered a good time to leave to pursue other interests, but many stayed on to continue flying.

"I think it was a very sobering time," Jerry Ross said. "I mean, we always knew that that was a possibility, that we would have such a catastrophic accident. I think there was a lot of frustration that we found out fairly soon afterwards that the finger was being pointed at the joint design and, in fact, that there had been quite a bit of evidence prior to the accident that that joint design was not totally satisfactory. Most of us had never heard that. We were very shocked, disappointed, mad."

Several astronauts left the office relatively soon after the accident, Ross said, and a stream of others continued to leave for a while afterward. "There were more people that left for various different reasons; some of them for frustration, some of them knowing that the preparations for flight were going to take longer, some of them responding to spouses' dictates or requests that they leave the program now that they'd actually had an accident."

Ross discussed his future at NASA with his family and admitted to having some uncertainty at first as to what he wanted to do.

I had mixed feelings at first about wanting to continue to take the risk of flying in space, but at the same time, all of the NASA crew members on the shuttle were good friends of mine, and I felt that if I were to quit and everybody else were to quit, then they would have lost their lives for no good benefit or progress. If you reflect back on history, any great undertaking has had losses; you know, wagon trains going across the plains or ships coming across the ocean. I was just watching a TV show that said in the 1800s, one out of every six ships that went across . . . the ocean from Europe to here didn't make it. So there's risk involved in any type of new endeavor that's going on. And I got into the program with my eyes wide open, both for the excitement and the adventure of it, but also I felt very strongly that it was important that we do those kinds of things for the future of mankind and for the good of America.

After getting through the shock and getting through the memorial services and all that, even though I was frustrated I wasn't getting a whole lot to do to help with the recovery effort, I was very determined that I was not going to leave, after talking with the family and getting their agreement, and that I was going to do whatever I could to help us get back to flying as soon as we could and to do it safer.

Sources

Books

Allen, Joseph P. *Entering Space: An Astronaut's Odyssey*. New York: Stewart, Tabori and Chang, 1984.

Cooper, Henry S., Jr. *Before Lift-Off: The Making of a Space Shuttle Crew*. Baltimore: John Hopkins University Press, 1987.

Evans, Ben. *Columbia: Her Missions and Crews*. Germany: Springer Praxis, 2005.

Froehlich, Walter. *Spacelab: An International Short-Stay Orbiting Laboratory*. Washington DC: NASA, 1983.

Grey, Jerry. *Enterprise*. New York: William Morrow, 1979.

Harland, David Michael. *The Story of the Space Shuttle*. Chichester UK: Springer Praxis, 2004.

Heppenheimer, T. A. *The Space Shuttle Decision: NASA's Search for a Reusable Space Vehicle*. Washington DC: NASA History Office, 1999.

Hitt, David, Owen Garriott, and Joe Kerwin. *Homesteading Space: The Skylab Story*. Lincoln: University of Nebraska Press, 2008.

Jenkins, Dennis R. *Space Shuttle: The History of the National Space Transportation System; The First 100 Missions*. Cape Canaveral FL: Dennis Jenkins, 2004.

Joels, Kerry Mark, and Gregory P. Kennedy. *The Space Shuttle Operator's Manual*. New York: Ballantine Books, 1982.

Lord, Douglas R. *Spacelab: An International Success Story*. Washington DC: NASA Scientific and Technical Information Division, 1987.

Marshall Space Flight Center. *Science in Orbit: The Shuttle and Spacelab Experience, 1981–1986*. Washington DC: NASA, 1988.

Mullane, Mike. *Riding Rockets: The Outrageous Tales of a Space Shuttle Astronaut*. New York: Scribner, 2006.

Nelson, Bill. *Mission: An American Congressman's Voyage to Space*. San Diego: Harcourt Brace Jovanovich, 1988.

Shayler, David J., and Colin Burgess. *NASA's Scientist-Astronauts*. Chichester UK: Springer Praxis, 2007.

Vaughan, Diane. *The Challenger Launch Decision: Risky Technology, Culture, and Deviance at NASA*. Chicago: University of Chicago Press, 1996.

Interviews and Personal Communications

Allen, Joseph P. Interviewed by Jennifer Ross-Nazzal. Houston, 28 January 2003.

Allen, Joseph P. Interviewed by Jennifer Ross-Nazzal. Washington DC, 16 March 2004.

Allen, Joseph P. Interviewed by Jennifer Ross-Nazzal. Washington DC, 18 March 2004.

Allen, Joseph P. Interviewed by Jennifer Ross-Nazzal. Washington DC, 18 November 2004.

Allen, Joseph P. Interviewed by Jennifer Ross-Nazzal. McLean VA, 18 April 2006.

Bobko, Karol J. "Bo." Interviewed by Summer Chick Bergen. Houston, 12 February 2002.

Bolden, Charles F. Interviewed by Sandra Johnson. Houston, 6 January 2004.

Bolden, Charles F. Interviewed by Sandra Johnson. Houston, 15 January 2004.

Brand, Vance D. Interviewed by Rebecca Wright. Houston, 25 July 2000.

Brand, Vance D. Interviewed by Rebecca Wright. Houston, 12 April 2002.

Brandenstein, Daniel C. Interviewed by Carol Butler. Kirkland WA, 19 January 1999.

Cleave, Mary L. Interviewed by Rebecca Wright. Washington DC, 5 March 2002.

Covey, Richard O. Interviewed by Jennifer Ross-Nazzal. Houston, 1 November 2006.

Covey, Richard O. Interviewed by Jennifer Ross-Nazzal. Houston, 15 November 2006.

Covey, Richard O. Interviewed by Jennifer Ross-Nazzal. Houston, 7 February 2007.

Covey, Richard O. Interviewed by Jennifer Ross-Nazzal. Houston, 28 March 2007.

Crippen, Robert L. Interviewed by Rebecca Wright. Houston, 26 May 2006.

Dunbar, Bonnie J. Interviewed by Jennifer Ross-Nazzal. Houston, 22 December 2004.

Dunbar, Bonnie J. Interviewed by Jennifer Ross-Nazzal. Houston, 20 January 2005.

Dunbar, Bonnie J. Interviewed by Jennifer Ross-Nazzal. Houston, 23 March 2005.

Dunbar, Bonnie J. Interviewed by Jennifer Ross-Nazzal. Houston, 14 September 2005.

Engle, Joe H. Interviewed by Rebecca Wright. Houston, 22 April 2004.

Engle, Joe H. Interviewed by Rebecca Wright. Houston, 5 May 2004.

Engle, Joe H. Interviewed by Rebecca Wright. Houston, 27 May 2004.

Engle, Joe H. Interviewed by Rebecca Wright. Houston, 3 June 2004.

Engle, Joe H. Interviewed by Rebecca Wright. Houston, 24 June 2004.

Fabian, John M. Interviewed by Jennifer Ross-Nazzal. Houston, 10 February 2006.

Fullerton, C. Gordon. Interviewed by Rebecca Wright. NASA Dryden Flight Research Center, Edwards CA, 6 May 2002.

Garriott, Owen K. Interviewed by Kevin M. Rusnak. Houston, 6 November 2000.

Gregory, Frederick D. Interviewed by Rebecca Wright. Washington DC, 29 April 2004.

Gregory, Frederick D. Interviewed by Jennifer Ross-Nazzal. Washington DC, 14 March 2005.

Gregory, Frederick D. Interviewed by Rebecca Wright. Annapolis MD, 18 April 2006.

Haise, Fred W. Interviewed by Doug Ward. Houston, 23 March 1999.

Hart, Terry J. Interviewed by Jennifer Ross-Nazzal. Houston, 10 April 2003.

Hartsfield, Henry W. "Hank," Jr. Interviewed by Carol Butler. Houston, 12 June 2001.

Hartsfield, Henry W. "Hank," Jr. Interviewed by Carol Butler. Houston, 15 June 2001.

Hauck, Frederick H. Interviewed by Jennifer Ross-Nazzal. Bethesda MD, 10 November 2003.

Hauck, Frederick H. Interviewed by Jennifer Ross-Nazzal. Bethesda MD, 17 March 2004.

Hawley, Steven A. Interviewed by Sandra Johnson. Houston, 4 December 2002.

Hawley, Steven A. Interviewed by Sandra Johnson. Houston, 17 December 2002.

Hawley, Steven A. Interviewed by Sandra Johnson. Houston, 14 January 2003.

Leestma, David C. Interviewed by Jennifer Ross-Nazzal. Houston, 26 November 2002.

Mattingly, T. K. Interviewed by Kevin M. Rusnak. Houston, 22 April 2001.

Mattingly, T. K. Interviewed by Rebecca Wright. Costa Mesa CA, 6 November 2001.

Mullane, Richard M. Interviewed by Rebecca Wright. Albuquerque, 24 January 2003.

Nelson, George D. "Pinky." Interviewed by Jennifer Ross-Nazzal. Bellingham WA, 6 May 2004.

O'Connor, Bryan. Interviewed by Sandra Johnson. Washington DC, 17 March 2004.

O'Connor, Bryan. Interviewed by Sandra Johnson. Washington DC, 20 April 2006.

O'Connor, Bryan. Interviewed by Sandra Johnson. Washington DC, 18 December 2006.

Peterson, Donald H. Interviewed by Jennifer Ross-Nazzal. Houston, 14 November 2002.

Ride, Sally K. Interviewed by Rebecca Wright. San Diego, 22 October 2002.

Ride, Sally K. Interviewed by Rebecca Wright. San Antonio TX, 6 December 2002.

Ross, Jerry L. Interviewed by Jennifer Ross-Nazzal. Houston, 4 December 2003.

Ross, Jerry L. Interviewed by Jennifer Ross-Nazzal. Houston, 26 January 2004.

Ross, Jerry L. Interviewed by Jennifer Ross-Nazzal. Houston, 5 February 2004.

Shaw, Brewster H., Jr. Interviewed by Kevin M. Rusnak. Houston, 19 April 2002.

Shriver, Loren J. Interviewed by Rebecca Wright. Houston, 16 December 2002.

Thagard, Norman E. Interviewed by Rich Dinkel. Tallahassee FL, 23 April 1998.

Walker, Charles D. Interviewed by Jennifer Ross-Nazzal. Washington DC, 19 November 2004.

Walker, Charles D. Interviewed by Jennifer Ross-Nazzal. Washington DC, 17 March 2005.

Walker, Charles D. Interviewed by Sandra Johnson. Houston, 14 April 2005.
Walker, Charles D. Interviewed by Sandra Johnson. Springfield VA, 7 November 2006.
Weitz, Paul J. Interviewed by Rebecca Wright. Flagstaff AZ, 26 March 2000.
Weitz, Paul J. Interviewed by Carol Butler. Houston, 8 November 2000.

Other Sources

NASA. http://www.nasa.gov.
collectSPACE. http://www.collectspace.com.
Space Shuttle Press Kits. Johnson Space Center, NASA. http://www.jsc.nasa.gov/history/shuttle_pk/shuttle_press.htm.
Report of the Presidential Commission on the Space Shuttle "Challenger" Accident. Washington DC, 1986.
"Skylab Mission Overview." *NASA Food Technology Commercial Space Center Newsletter* 7, no. 1 (February 2006): 16.

Index

Page numbers in italics refer to illustrations.

In the Outward Odyssey: A People's History of Spaceflight series

Into That Silent Sea: Trailblazers of the Space Era, 1961–1965
Francis French and Colin Burgess
Foreword by Paul Haney

In the Shadow of the Moon: A Challenging Journey to Tranquility, 1965–1969
Francis French and Colin Burgess
Foreword by Walter Cunningham

To a Distant Day: The Rocket Pioneers
Chris Gainor
Foreword by Alfred Worden

Homesteading Space: The Skylab Story
David Hitt, Owen Garriott, and Joe Kerwin
Foreword by Homer Hickam

Ambassadors from Earth: Pioneering Explorations with Unmanned Spacecraft
Jay Gallentine

Footprints in the Dust: The Epic Voyages of Apollo, 1969–1975
Edited by Colin Burgess
Foreword by Richard F. Gordon

Realizing Tomorrow: The Path to Private Spaceflight
Chris Dubbs and Emeline Paat-Dahlstrom
Foreword by Charles D. Walker

The X-15 Rocket Plane: Flying the First Wings into Space
Michelle Evans
Foreword by Joe H. Engle

Wheels Stop: The Tragedies and Triumphs of the Space Shuttle Program, 1986–2011
Rick Houston
Foreword by Jerry Ross

Bold They Rise: The Space Shuttle Early Years, 1972–1986
David Hitt and Heather R. Smith
Foreword by Bob Crippen